Alicyclic Chemistry

V. K. Ahluwalia · Renu Aggarwal

Alicyclic Chemistry

Second Edition

Ane Books
Pvt. Ltd.

Springer

V. K. Ahluwalia
Department of Chemistry
University of Delhi
Delhi, India

Renu Aggarwal
Department of Chemistry, Gargi College
University of Delhi
Delhi, India

ISBN 978-3-031-36067-1   ISBN 978-3-031-36068-8   (eBook)
https://doi.org/10.1007/978-3-031-36068-8

Jointly published with Ane Books Pvt Ltd.
The print edition is not for sale in South Asia (India, Pakistan, Sri Lanka, Bangladesh, Nepal and Bhutan) and Africa. Customers from South Asia and Africa can please order the print book from: ANE Books Pvt. Ltd.
ISBN of the Co-Publisher's country edition: 978-93-8921-269-3.

This Springer imprint is published by the registered company Springer Nature Switzerland AG
The registered company address is: Gewerbestrasse 11, 6330 Cham, Switzerland

Preface to First Edition

Alicyclic Chemistry is the chemistry of compounds, which contains one or more carbocyclic rings containing three or more number of carbon atoms. The present book describes the synthesis and properties of such compounds which includes cyclopropane, cyclobutane, cyclopentane, cyclohexane, cycloheptane and cycloheptatriene. The conformation of cycloalkanes has been described in detail. Beside these, cyclooctane and its derivatives, civetone and muscone have also been discussed.

The book also deals with the non-benzenoid aromatic compounds including annulenes, metallocenes and azulenes. Bridged ring compounds particularly adamantane and twistane have also been incorporated. Cage molecules like cubane and prismane have been described.

The book also contains discussions on tropone, tropolones, fluxional molecules, catenanes and rotaxanes.

It is hope that the book will be useful to undergraduate and postgraduate students in organic chemistry.

The authors express their thanks to Prof. J.M. Khanna for his help and suggestions.

– **Authors**

Preface to Second Edition

The book 'Alicyclic Chemistry' has been completely revised and updated. A new secton on cycloalkanes containing one hetero atom (Heterocyclic compounds) has been incorporated. The mechanism of all reactions have been included.

At the end, questions including multiple choice questions, fill in the blanks and short answer-questions have been included. It is hoped that the revised addition will be helpful for preparing the students for varios compentitive examinations.

Any suggestions from the users of this book will be greatifully acknowledged and included in subsequent addition.

– **Authors**

Contents

About the Authors

Dr. V. K. Ahluwalia was Professor of Chemistry at Delhi University for more than three decades teaching graduate, postgraduate and M.Phil. students. He was also Postdoctoral Fellow between 1960 and 1962 and worked with renowned global names from prestigious international universities. He was Visiting Professor of Biomedical Research at University of Delhi. V.K. Ahluwalia was widely regarded as Leading Subject Expert in chemistry and allied subjects along with being a "Choice Award for an Outstanding Academic Title" winner. He had published more than 100 titles. Apart from books, he had published more than 250 research papers in national and international journals.

Renu Aggarwal is working as Associate Professor in the Chemistry Department at Gargi College, New Delhi, for the last thirty years.

Introduction

Cycloalkanes, as the name implies are obtained from alkanes by removing one hydrogen atom from each of the two terminal methyl groups.

Thus, the molecular formula of cycloalkane is C_nH_{2n} compared to alkanes which have the molecular formula C_nH_{2n+2}.

The cycloalkanes are referred to as carbocyclic compounds. Cycloalkanes having medium size rings (C_5 or C_6) are qutite stable and resemble alkanes. However, compounds having rings with carbon atoms less than five are quite interesting, since their chemical properties are intermediate between alkanes and alkenes.

❑❑❑

© The Author(s) 2023
V. K. Ahluwalia and R. Aggarwal, *Alicyclic Chemistry*,
https://doi.org/10.1007/978-3-031-36068-8_1

2

Nomenclature of Cycloalkanes

2.1 MONOCYCLIC COMPOUNDS

These are cycloalkanes having only one ring and are named by attaching the prefix cyclo to the name of the alkanes having the same number of carbon atoms. Some examples are given below:

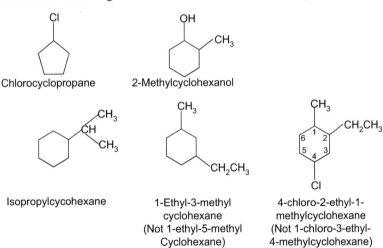

Cyclopropane Cycoobutane Cyclopentane Cyclohexane

In case of substituted cycloalkanes, the nomenclature is straight forward. These are named as alkylcycloalkanes, halocycloalkanes and so on. In case, only one substituent is present, it is not necessary to designate its position. In case two substituents are present, the ring is numbered starting from the substitution first in the alphabet and the

Chlorocyclopropane

2-Methylcyclohexanol

Isopropylcycohexane

1-Ethyl-3-methyl cyclohexane
(Not 1-ethyl-5-methyl Cyclohexane)

4-chloro-2-ethyl-1-methylcyclohexane
(Not 1-chloro-3-ethyl-4-methylcyclohexane)

© The Author(s) 2023
V. K. Ahluwalia and R. Aggarwal, *Alicyclic Chemistry*,
https://doi.org/10.1007/978-3-031-36068-8_2

numbering is done so that the next substituent gets the lower number possible. However in case three or more substituents are present the numbering is started at the substitutent that leads to the lowest set of locants.

In case a single ring system is attached to a single chain with greater number of carbon atoms (than in the cycloalkane) or when more than one ring system is attached is a single chain, in such cases the compounds are named as cycloalkylalkans. Some examples are given below:

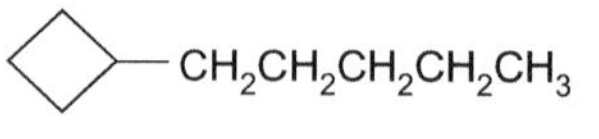

1-cyclobutyl pentane 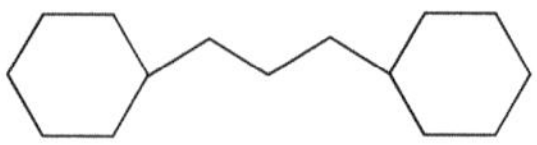   1, 3-Dicyclohexylpropane

Cycloalkanes containing one double bond are named as cycloalkenes. Some examples are given below:

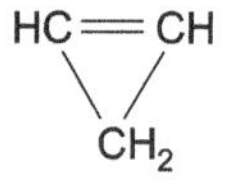

Cyclopropene 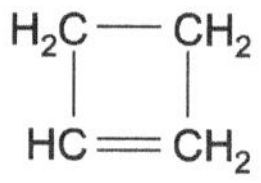   Cyclobutene 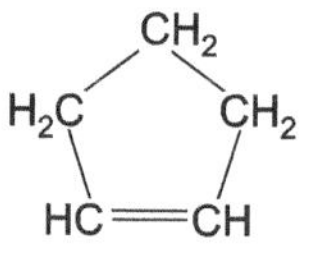   Cyclopentene 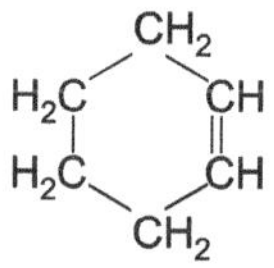   Cyclohexene

If the cycloalkanes contain more than one double bond, the location of the olefinic bonds may be obvious as in the case of the following three cycloalkenes, in which only one structure is possible.

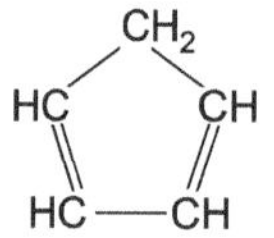

Cycoopentadiene 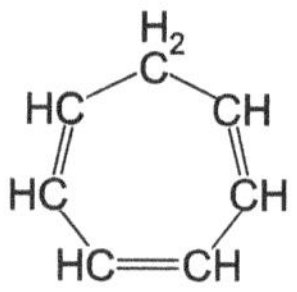   Cycooheptatriene 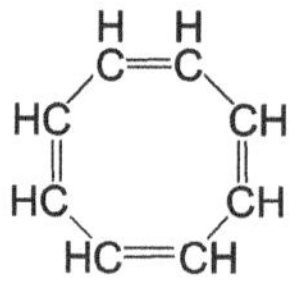   Cyclooctatetraene

The substitued cycloalkanes are named as per IUPAC system of nomencluture as given below:

(*i*) The cycloalkenes are numbered in a way that gives the carbon atoms of the double bond 1 and 2 positions and that also the substituent groups the lower numbers at the first point of difference. With substituted cycloalkenes it is not necessary to specify the position of the double bond since it always begins with C_1 and C_2. Some examples are given below:

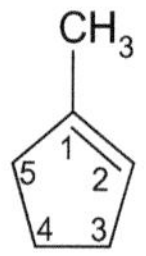

1-Methylcyclopentene   3, 5-Dimethylchlorocyclohexene
(Not 2-Methylcyclopentene)   (Not 4, 6-Dimethylcyclohexene)

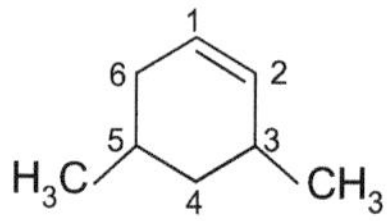

(*ii*) Cycloalkenes containing an alcohol group are named as cycloalkenols, the alcoholic carbon is given the lowest number.

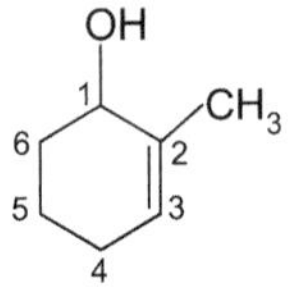

2-Methyl-2-cyclonexen-1-ol

2.2 BICYCLIC COMPOUNDS

The presence of more than one ring is indicated by the prefix *bicyclo-*, *tricyclo-*, etc. Such compounds are known as polycyclic compounds. They contain two or more rings having two or more carbon atoms in common. In such cases, the ring carbons are numbered commencing from any one of the carbon atoms at the junction (the bridge head carbon) and is continued to the next junction (bridge head carbon) first along the longest path (maximum number of carbon atoms), and then along the shorter and shorter paths, successively. The number of carbon atoms involved in each of these paths is written in square brackets in descending order and inserted in between the words bicyclo and alkane. Following examples illustrate the nomenclatre of bicyclic compounds:

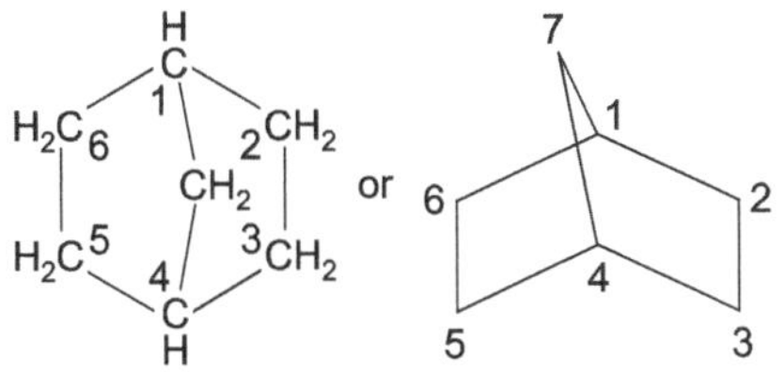

Bicyclo [2.2.1] heptane

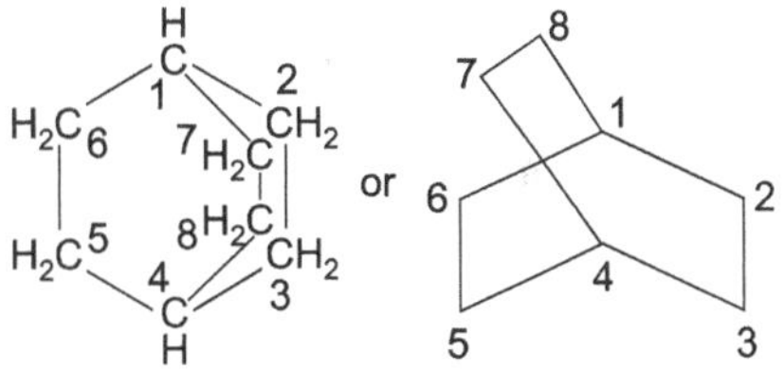

Bicyclo [2.2.2] octane

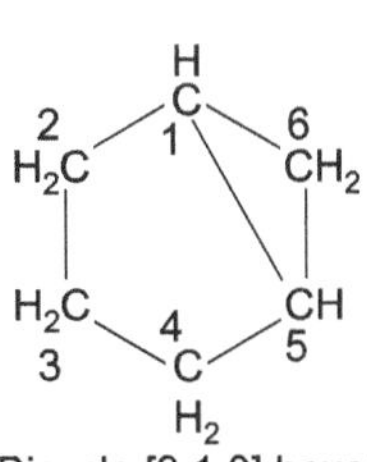

Bicyclo [3.1.0] hexane

Recently, certain polycyclic hydrocarbons having fascinating structures have been synthesised. These are given common names, indicating their shape. Two such examples are cubane (shape of a cube) and baskatene (shape of a basket).

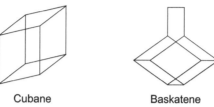

Cubane   Baskatene

❏❏❏

3

Synthesis of Cycloalkanes

1.3 SYNTHESIS OF CYCLOALKANES

Cycloalkanes are mostly synthesised from aliphatic compounds by using suitable reactions to bring about cyclisation. In most of the methods used the products obtained are derivatives ol cycloalkanes, which can be converted into cycloalkanes by using appropriate reactions. Cycloalknes are also prepared from aromatic compounds. Following are given some of the common methods used for the synthesis of cycloalkanes.

(*i*) **From Aromatic Compounds**

Aromatic compounds, e.g. benzene on catalytic hydrogenation yield the corresponding saturated compounds (alicyclic compounds).

Benzene + 3H$_2$ $\xrightarrow{\text{Catalyst}}$ Cyclohexane

Cyclohexane can also be obtained from phenol. Thus, phenol on reduction gives cyclohexanol, which on dehydration followed by hydrogenation of the formed intermediate (cyclohexene) gives cyclohexane.

Phenol + 3H$_2$ $\xrightarrow{\text{Catalyst}}$ Cyclohexanol $\xrightarrow[-\text{H}_2\text{O}]{\text{H}^+,\ \Delta}$ Cyclohexene $\xrightarrow{\text{H}_2\text{- catalyst}}$ Cyclohexane

(*ii*) **From α, ω-dihaloalknes**

Treatment of α, ω-dihalogen derivatives of alkanes (e.g., 1, 3-dibromopropane or 1, 6-dibromo hexane) on treatment with sodium or zinc give the corresponding cycloalkane.

© The Author(s) 2023
V. K. Ahluwalia and R. Aggarwal, *Alicyclic Chemistry*,
https://doi.org/10.1007/978-3-031-36068-8_3

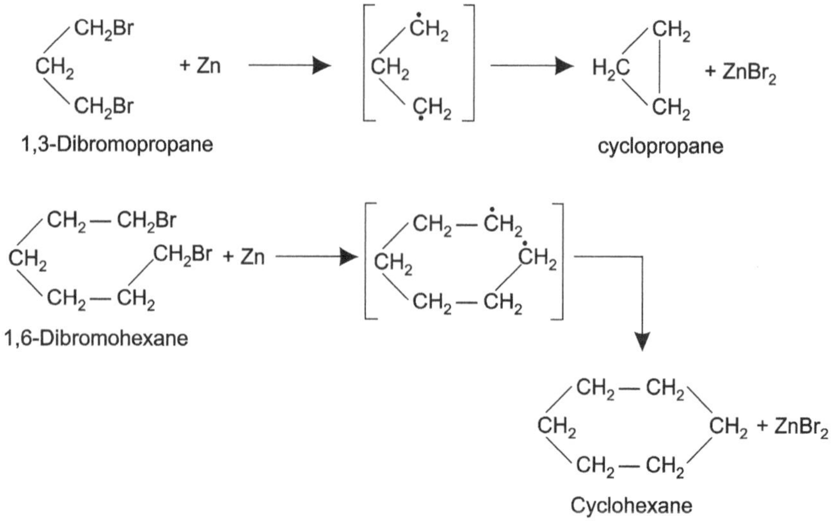

1,3-Dibromopropane   cyclopropane

1,6-Dibromohexane

Cyclohexane

The above reaction can be regarded as intramolecular Wuntz reaction. In this reaction if in the dihalide, $X — (CH_2)_n — X$, $n > 6$, the products of interomolecular reaction are obtained.

(iii) From Diethylmalonate

The reaction of diethylmalonate with α, ω-dihalides (viz, 1, 2-, 1, 3- or 1, 4-dibromoalkanes in presence of sodium ethoxide give cycloalkane carboxylic acids (viz, cyclopropane carboxylic acids, cyclobutane carboxylic acid or cyclopentane carboxylic acid respectively). These carboxylic acids on decarboxylation give the corresponding cycloalkane.

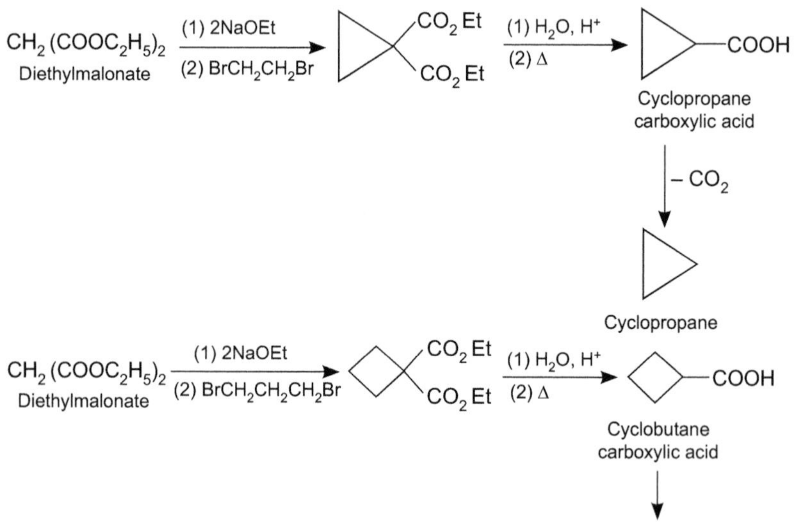

Cyclopropane carboxylic acid

Cyclopropane

Cyclobutane carboxylic acid

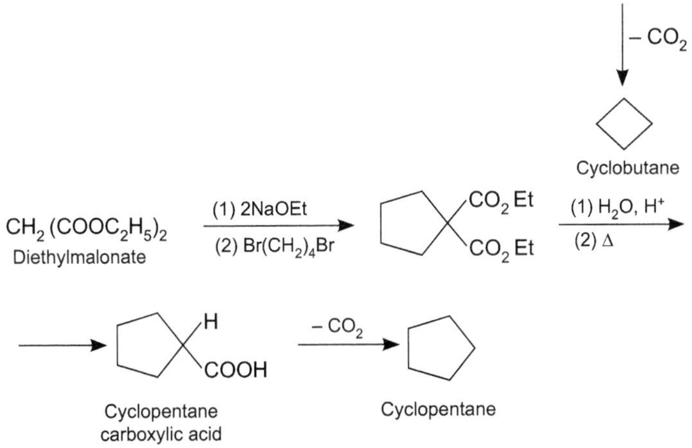

$$\text{CH}_2(\text{COOC}_2\text{H}_5)_2 \xrightarrow[\text{(2) Br(CH}_2)_4\text{Br}]{\text{(1) 2NaOEt}} \quad \begin{array}{c} \text{CO}_2\text{Et} \\ \text{CO}_2\text{Et} \end{array} \xrightarrow[\text{(2) } \Delta]{\text{(1) H}_2\text{O, H}^+}$$

Diethylmalonate

$$\downarrow - \text{CO}_2$$

Cyclobutane

$$\longrightarrow \begin{array}{c} \text{H} \\ \text{COOH} \end{array} \xrightarrow{- \text{CO}_2} $$

Cyclopentane
carboxylic acid   Cyclopentane

Cyclobutane 1, 2-dicarboxylic acid can also be obtained from diethylmalonate as given below:

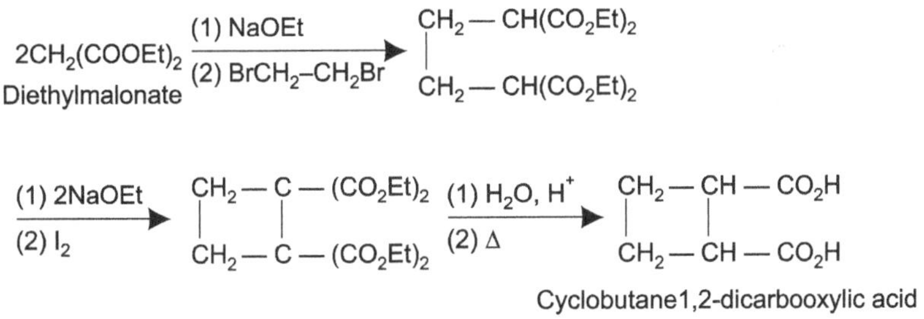

$$2\text{CH}_2(\text{COOEt})_2 \xrightarrow[\text{(2) BrCH}_2-\text{CH}_2\text{Br}]{\text{(1) NaOEt}} \quad \begin{array}{c} \text{CH}_2 - \text{CH}(\text{CO}_2\text{Et})_2 \\ | \\ \text{CH}_2 - \text{CH}(\text{CO}_2\text{Et})_2 \end{array}$$

Diethylmalonate

$$\xrightarrow[\text{(2) I}_2]{\text{(1) 2NaOEt}} \quad \begin{array}{c} \text{CH}_2 - \text{C} - (\text{CO}_2\text{Et})_2 \\ | \qquad | \\ \text{CH}_2 - \text{C} - (\text{CO}_2\text{Et})_2 \end{array} \xrightarrow[\text{(2) } \Delta]{\text{(1) H}_2\text{O, H}^+} \quad \begin{array}{c} \text{CH}_2 - \text{CH} - \text{CO}_2\text{H} \\ | \qquad | \\ \text{CH}_2 - \text{CH} - \text{CO}_2\text{H} \end{array}$$

Cyclobutane1,2-dicarbooxylic acid

The reaction of ethyl acetoacetate with 1, 5-dibromobutane in presence of sodium ethoxide gives cyclohexylmethyl ketone as shown below:

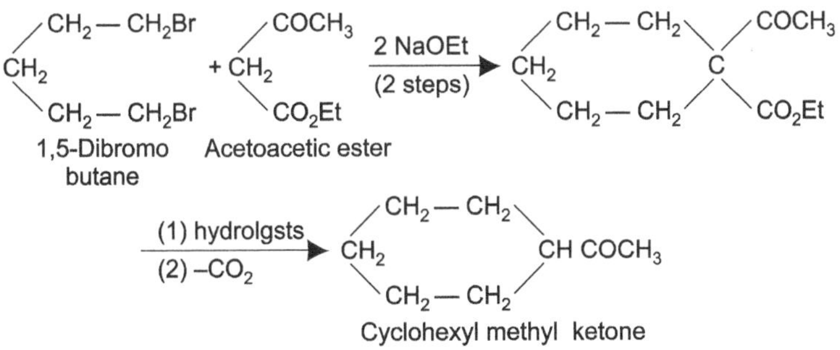

$$\begin{array}{c} \text{CH}_2 - \text{CH}_2\text{Br} \\ \text{CH}_2 \\ \text{CH}_2 - \text{CH}_2\text{Br} \end{array} + \begin{array}{c} \text{COCH}_3 \\ \text{CH}_2 \\ \text{CO}_2\text{Et} \end{array} \xrightarrow[\text{(2 steps)}]{\text{2 NaOEt}} \begin{array}{c} \text{CH}_2 - \text{CH}_2 \\ \text{CH}_2 \qquad \text{C} \\ \text{CH}_2 - \text{CH}_2 \end{array} \begin{array}{c} \text{COCH}_3 \\ \text{CO}_2\text{Et} \end{array}$$

1,5-Dibromo   Acetoacetic ester
butane

$$\xrightarrow[\text{(2) } -\text{CO}_2]{\text{(1) hydrolgsts}} \begin{array}{c} \text{CH}_2 - \text{CH}_2 \\ \text{CH}_2 \qquad\qquad \text{CH COCH}_3 \\ \text{CH}_2 - \text{CH}_2 \end{array}$$

Cyclohexyl methyl ketone

(iv) From Dicarboxylic Acids

Calcium, barium or thorium salts of dicarboxylic acids yield cyclic ketones on heating. For example,

Barium adipate   Cyclopentanone

The above method can also be used for the synthesis of macrocyclic rings (having as many as 34 carbon atoms) by the distillation of thorium, cerium or yttricum salts of dicarboxylic acid mixed with copper powder at 300°C in vacuo.

As per **Blanc's rule**, this method in useful for the synthesis of 5- and 6-membered cyclic ketones. Blanc's rule is an empirical rule and is based on experimental observation of the case of formation of cyclic ketones and anhydrides. The Blanc's rule states that pyrolysis of 1,4- and 1,5-dicarboxylic acids yield cyclic anhydrides white 1, 6- and 1, 7-discarboxylic acids yield cyclic ketones and higher dicarboxylic acids remain uneffected.

The cyclic ketones obtained by the above method could be converted into cycloalkanes by any of the following methods.

Cyclopentanone   Cyclopentane

(v) **Dieckmann Cyclisation**

Diesters of C_6 and C_7 dibasic acids undergo an **intramolecular Claisen condensation** in presence of base to give good yields of cyclic β-ketoesters. This is known as Dieckmann condensation. It is of considerable value in the synthesis of cyclic compounds. For example, ethyl esters of adipic acid and pimelic acids give 2-carbethoxycyclo pentanone and 2-carbethoxycyclohexanone resp.

CH$_2$-CH$_2$-COOEt
|
CH$_2$-CH$_2$-COOEt

Ethyl adipate

→ Na Or NaOEt →

2-Carbethoxycyclopentanone

CH$_2$-CH$_2$-COOEt
/
CH$_2$
\
CH$_2$-CH$_2$-COOEt

Ethyl pimelate

→ Na Or NaOEt →

2-Carbethoxycyclohexanone

Dieckmann condensation best proceeds with esters having 6, 7 or 8 carbon atoms and gives stable rings with 5, 6 or 7 carbons.

The mechanism of Dieckmann condensation involves abstraction of a proton from one of the α-carbon atoms to give a carbanion, which attacks the carbonyl carbon of the other ester group. Finally, the β-ketoester is formed by expulsion of ethoxide anion ($^-$OEt).

Ethyl adipate

2-Carbethoxycyclopentanone

It is found that Dieckmann condensation proceeds very well on sonication in a short time. On sonication, potassium is easily transformed to a silver blue suspension in toluene. The ultrasonically dispersed potassium is extremely helpful in Dieckmann condensation (cyclisation).

EtO$_2$C(CH$_2$)$_4$CO$_2$Et
Diethyladipate

→ K,)))) Toluene 5 min →

2-Carbethoxycyclopentanone

In the above condensation, bases like Bu^{t}OK, Bu^{t}ONa, EtOK or EtONa could also be used.

Esters lower than adipic acid form products by intermolecular condensation followed by cyclisation.

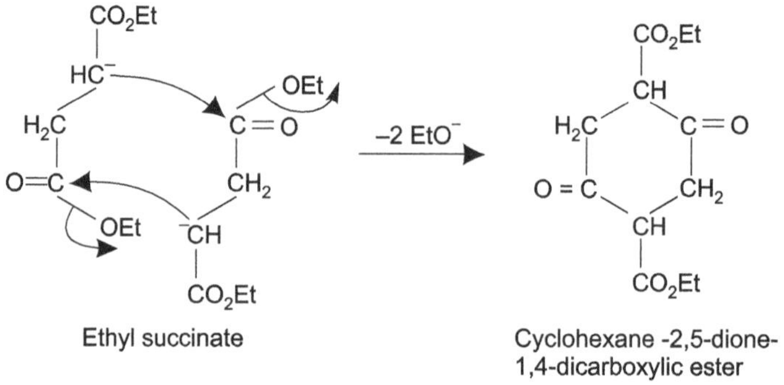

Ethyl succinate

Cyclohexane -2,5-dione-
1,4-dicarboxylic ester

Five membered ring compounds can be prepared by the intramolecular condensation between appropriate reactants. Thus, condensation of oxalic ester with glutaric acid ester gives the five membered cyclic product.

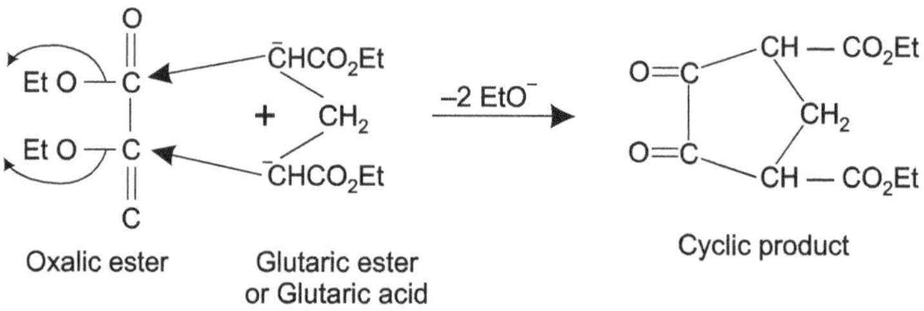

Oxalic ester   Glutaric ester
                   or Glutaric acid

Cyclic product

(vi) Thorpe-Ziegler's Method

In this method a dicyano compound NC $CH_2(CH_2)_2$ CH_2CN, reacts with $LiNEt_2$ in excess of benzene or toluene. The resulting cyclic imino compound on hydrolysis and subsequent decarboxylation gives the corresponding cyclic ketone is good yield.

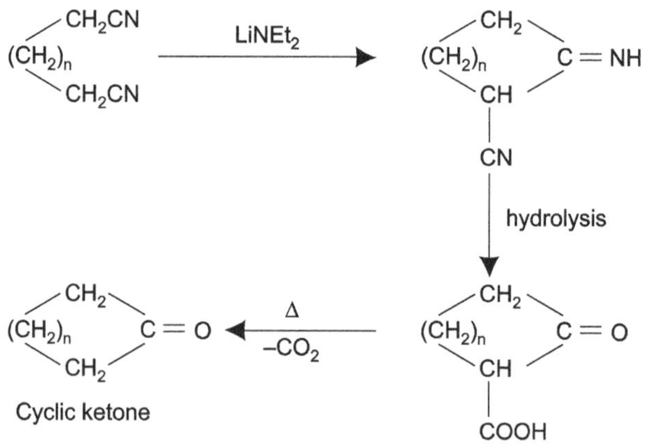

Cyclic ketone

Excess the solvent (benzene or toluene) is required to ensure intramolecular reaction instead of the intermolecular reaction. This is one the best methods for preparing macrocyclic compounds in good yield.

(*vii*) **Acyloin Condensation**

In this method diesters are refluxed with metallic sodium in aprotic solvents like benzene, ether, toluene or xylene, free from oxygen, to give cyclic α-hydroxy ketones, called cyclic acyloins.

Diester

Cyclic acyloin,
where n = 10 - 20 or more

The mechanism of the reaction is as follows:

To account for the ready formation of large rings, it is believed that the two ends of the esters are adsorbed weakly on the surface of sodium metal. In this way, the reactive ends are not available for intermolecular reaction to complete with cyclisation.

Solution surface

(vi) Cycloaddition Reactions

These reactions involve addition between two unsaturated molecules resulting in the formation of a cyclic product. In these reactions, π electrons of the two molecules form two σ bonds and are designated by writing the number of π electrons of the two molecules in square brackets. For example [2 + 2] cycloaddition refers to the addition of two alkene molecules using two π electrons each to form a cyclobutane ring. Similarly [4 + 2] cycloaddition refer to the addition of a molecule of an alkene (2π elections) to a conjugate diene (4π electron), resulting in the formation cyclohexane derivative. Some of the important cycloaddition reaction are discussed below:

(a) [4 + 2] cycloaddition. Example: Diels' Alder reaction.

In this type of cycloaddition reaction a conjugated diene (4π electron) is reacted with an alkene (2π electron), called a dienophile, to form a cyclohexene derivative.

| Diene | Dienophile | Cyclohexene derivative |

The reaction can take place even at room temperature if the dienophile has an electron attracting groups like carbonyl, cyano, CO_2H, Cl etc. For example.

(b) [2 + 2] Cycloaddition

As already stated in this type of addition, two alkene molecules combine together to form a cyclobutane derivative.

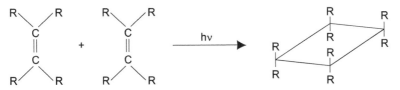

The above reaction proceed under photochemical conditions and are called photodimerisation. This reaction proceed in a concerted manner involving cyclic transition state.

Certain suitably substituted alkens like $R_2C = CF_2$ or $R_2C = CHX$ (X = $-COR$, $-CN$, $-COOR$ etc.) form cyclobutane derivatives under thermal conditions. These reaction proceed via a free radical mechanism involving diradical intermediates.

$$\underset{CR_2}{\overset{CF_2}{\|}} + \underset{CR_2}{\overset{CR_2}{\|}} \xrightarrow{\Delta} \underset{F_2C - \overset{\bullet}{C}R_2}{\overset{F_2C - CR_2}{|}} \longrightarrow \underset{F_2C - CR_2}{\overset{F_2C - CR_2}{|\quad\;|}}$$

Diradical

$$\underset{CR_2}{\overset{CF_2}{\|}} + \underset{CH - C_6H_5}{\overset{CH_2}{\|}} \xrightarrow{\Delta} \underset{R_2C\cdot\quad\cdot CH - C_6H_5}{\overset{F_2C - CH_2}{|}} \longrightarrow \underset{R_2C - CH - C_6H_5}{\overset{F_2C - CH_2}{|\quad\;|}}$$

R = Cl

(c) Addition of Carbene to alkenes

Highly reactive bivalent carbon species, carbenes, add to alkene molecule forming cyclopropane derivatives. The required carbene is generated *in situ* by any of the known methods. For example, the reaction of 2-butene and diazomethane on irradiation gives 1, 2-dimethylcyclopropane.

$$CH_3-CH= CH - CH_3 + :CH_2 \xrightarrow[CH_2 N_2]{h\nu} \underset{\underset{CH_2}{\diagup\diagdown}}{CH_3 - CH - CH - CH_3} + N_2$$

2-Butene   1,2-Dimethyl cyclopropane

◻◻◻

Properties of Cycloalkanes

4.1 PHYSICAL PROPERTIES

With a few exceptions, the boiling points and melting points of cycloalkanes show some gradation. A comparison of the melting and boiling points of cycloalkanes and alkanes (Table 1) show that the cycloalkanes melt and boil at higher temperature than the alkane. This may probably due to the fact that the zig-zag alkane molecules are constantly fluttering making it difficult for the molecules, in the solid state to be closely packed there by lowering their melting points as compared to cycloalkanes where the more rigid structure prevents this fluttering to a great extent. Also the intermolecular attractions in alkanes are less than those in corresponding cycloalkanes hence the lower boiling points of alkans.

Table 4.1 B.P's and M.P's of cycloalkanes of alkane

Cycloalkanes	B.P (°C)	M.P (°C)	Alkanes	B.P (°C)	M.P (°C)
Cyclopropane	− 33	−126.6	Propane	−42	−187
Cyclobutane	13	−80	n-Butane	0	−138
Cyclopentane	49	− 94	n-Pentane	36	−130
Cyclohexane	81	6.5	n-Hexane	69	− 95
Cycloheptane	118.5	− 12	n-Heptane	98	−90
Cyclooctane	149	13.5	n-octane	126	−57

Cycloalianes like alkanes are insoluble in water due to their low polarity and their inability to form hydrogen bonds. However, cycloalkanes are soluble in solvents of low polarity, like benzene, carbon tetrachloride. Also cycloalkanes (like alkanes) are lest dense of all groups of organic compounds. Their densities are considerably less than 1.00 g mol^{-1} the density of water at 4°C). So cycloalkanes float on water.

© The Author(s) 2023
V. K. Ahluwalia and R. Aggarwal, *Alicyclic Chemistry*,
https://doi.org/10.1007/978-3-031-36068-8_4

4.2 INFRARED SPECTRA

The infrared spectra of cycloalkanes are almost similar to those of alkanes and the absorption due to C–H stretching of the CH_2 groups falls in the same region except for strained molecules like cyclopropane, which show these absorption in the $3100 - 2940$ cm^{-1} region. Another band of moderate intensity, due to CH_2 scissoring appears in $1470 - 1440$ cm^{-1} region. The 1R spectra of cyclohexane is shown below as an representative example.

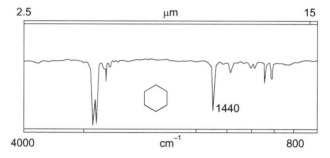

Fig. 4.1 IR Spectra of cyclohexane

4.3 ULTRAVIOLET SPECTRA

Like alkanes, cycloalkans are also transparent to UV light above 220 nm.

4.4 NUCLEAR MAGNETIC REASONAEE SPECTRA

The cycloalkans have two types of protons (axial and equatorial) having different environments. However, the nmr spectra of simple unsubstituted cycloalkanes show a sharp singlet at δ 1.5–2.0 (the region of methylene absorption), due to the fact that the two types of protons are indistinguishable

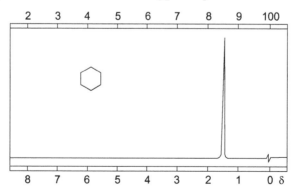

Fig. 4.2 NMR Spectra of cyclohexane

because of rapid reversible interconversions of one form of cyclohaxane into another. The NMR spectra of cyclohexane given earlier is quite illustrative.

However, at the temperature, when its equilibrium is frozen, two signals are observed — one *upfield* (due to *axial* protons) and the other downfield (due to *equatorial* protons)

4.5 CHEMICAL PROPERTIES OF CYCLOALKANES

The general formula, (C_nH_{2n}), of cycloalkanes suggest that they should be unsaturated like alkenes. However, they closely resemble alkanes in chemical properties. This similarity is attributed to the fact that all the carbon atoms of these compounds are sp^3 hybridized, forming sigma bond with two neighbouring carbons and two hydrogens. It has been found that cycloalkanes having 5 or more carbon atoms show high degree of stability and the compounds remain uneffected by acids, alkalies and common oxidizing agents like $KMnO_4$ under usual conditions. However, cyclopropane and cyclobutane are exceptions; they show a high degree of reactivity, cyclopropane being particularly more reactive.

The relative stability of cycloalkanes is discussed in a subsequent section. Following are given some of the common reactions of cyclopropanes.

(*i*) *Halogenation.* Halogenation of cyclopropane takes place in the presence of sunlight by free radical substitution.

(*ii*) *Addition reactions.* Cyclopropane, like alkenes undergoes addition reactions. Some typical examples are given below:

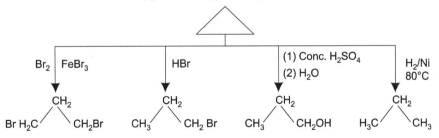

Cyclopropane, unlike alkene, is not hydroxylated with aq. $KMnO_4$ and also does not add ozone.

Cyclobutane is not as reactive as cyclopropane; it does not undergo addition reactions under ordinary conditions. However, it can be hydrogenated to n-butane at higher temperature.

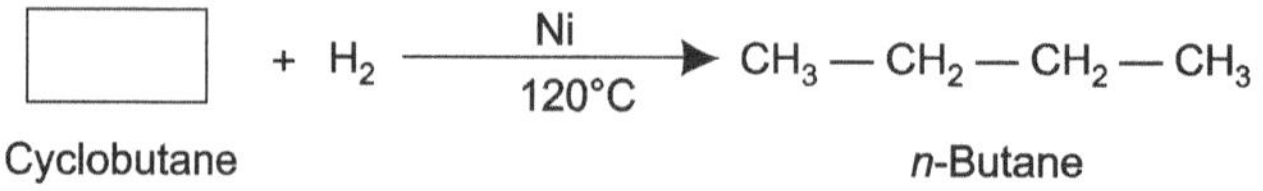

$$\boxed{} \quad + \text{ H}_2 \xrightarrow[\text{120°C}]{\text{Ni}} \text{CH}_3 - \text{CH}_2 - \text{CH}_2 - \text{CH}_3$$

Cyclobutane   *n*-Butane

4.5.1 Baeyers Strain Theory

It has already been stated (section 4.5) that cycloalkanes having 5 or more carbon atoms are more stable than cycloalkanes having 3 or 4 carbons. In order to explain this behaviours, Adolf von Baeyer (1885) proposed his well known Baeyer's strain theory. This theory is based purely on mechanical concept. Its main assumptions are as follows:

   (*i*) Carbon atom is tetrahedral in nature. All the four valencies are directed towards the four corners of a regular tetrahedron. The angle between any two bonds is 109.5°, which is known as tetrahedral angle. Any deviation from this angle would result in strain in the molecule. It can be seen that a double bond in alkenes causes considerable deviation from the tetrahedral angle, if we regard the angle between the two carbon-carbon valencies of a double bond to be zero. So the alkene molecule is highly strained and so alkenes or olefins are highly reactive.

  (*ii*) All carbon atoms comprising a cycloalkane ring are planar. This implies that cyclopropane ring is an equilateral triangle, cyclobutane ring is square and other cycloalkane rings are regular polygens.

 (*iii*) On the basis of the above mentioned shapes of cycloalkanes it is obvious that there is deviation from the normal tetrahedral angle causing thereby a lot of strain in the molecule. This strain (commonly referred to as an angle strain) depends upon the extent of deviation. This implies that greater the deviation of bond angle from 109.5°, the greater is the strain and so more unstable is the ring. In case of cyclopropane and cyclobutane, the bond angles must be compressed to 60° and 90°, respectively.

According to Baeyer, these deviations (listed in the table below) from normal tetrahedral angle can be calculated as follows:

Deviation from 109.5° for one bond (*d*) $= \dfrac{1}{2}$ (109.5°) − bond angle in the planar ring).

Table 4.2 Deviations from the normal tetrahedral angle in case of cycloalkanes and heats of combustion per CH_2 unit

Compound (geometry)	Bond angle in planar ring	Deviation	Heat of combustion per CH_2 group (kJ/mol)
Cyclopropane (equilateral triangle)	60°	$\frac{1}{2}(109.5 - 60) = 24.75°$	697.0
Cyclobutane (square planar)	90°	$\frac{1}{2}(109.5 - 90) = 9.75°$	686.0
Cyclopentane (regular pentagon)	108°	$\frac{1}{2}(109.5 - 108) = 0.75°$	664.0
Cyclohexane (regular hexagon)	120°	$\frac{1}{2}(109.5 - 120) = -5.25°$	658.7
Cycloheptane (regular heptagon)	128.5°	$\frac{1}{2}(109.5 - 128.5) = -9.5°$	662.4
Cyclooctane (regular octagon)	135°	$\frac{1}{2}(109.5 - 130) = 12.75°$	663.8

The positive values of deviation mean that the bond angles are compressed while negative values mean expansion of the bond angle from the tetrahedral angle. The molecule is strained in both the situations. As seen from the table, the deviation from the normal tetrahedral angle falls rapidly in going from C_3 to C_5 ring system and then there is slow rise. This is in agreement with the fact that cyclopropane, being more strained, is more reactive than cyclobutane. This is also in agreement with the observed facts.

4.5.2 Limitations of Baeyer's Strain Theory

1. According to Baeyer's strain theory, cyclopentane should be more stable than cyclohexane. This is not correct. In fact, cyclopentane ring can be opened at 300° but cyclohexane ring is very stable.

2. Cyclopropane ring is formed more readily than cyclobutane ring though it involves more strain than cyclobutane ring.

3. According to Baeyer's strain theory, compound containing more than six carbon atoms should be very unstable and should not exist because of greater strain in them. However, compounds up to C_{16} (civetone) and C_{17} (muscone) have been synthesized and are quite stable.

4. The deviations in cyclobutane and cycloheptane are nearly equal. So these are supposed to be equally unstable. However, it is found that cycloheptane is more stable than cyclobutane.

5. Heats of combustion per $-CH_2$ group of various cycloalkanes indicate that rings larger than cyclohexane are stable. However, according to Baeyer's strain theory higher rings with larger strain would be unstable.

Thus, the Baeyer's strain theory is well applicable to explain the reactivity and stability of cycloalkanes up to cyclobutane, but it does not apply for cyclohexane and higher cycloalkanes.

Determination of heats of combustion per –CH_2 group of cyclohexane is discussed in a separate section.

4.5.3 Sachse-Mohr Theory of Strainless Rings

It has already been stated that according to Baeyer strain theory, higher members of cycloalkanes with larger strain would be unstable. To explain this anamoly, Sachse (1890) proposed that, contrary to Baeyer's postulate, carbon atoms of cyclohexane and higher homologues are not planar but assume a strain-free puckered configuration, in which each carbon retains its normal valency angle of 109.5°. On this basis, Sachse could construct two models of cyclohexane called the boat and chair forms.

Boat form   Chair form

Initially Sachse's proposal was not acceptable as it was based on the assumption that two types of cylohexanes should exist and at that time only one type of cyclohexane was known. Subsequently, Mohr (1931) reconsidered Sachse's proposal and suggested the possibility of existence of two forms which readily undergo interconversion by rotation of single bonds involving negligible distortion of the tetrahedral angles. Only 5 kcal/mol energy is needed for their interconversion. This theory became known as *Sachse-Mohr theory of strainless rings*.

The two conformational isomers are in in dynamic equilibrium and have not been individually isolated. On the basis of theoretical and experimental evidence, it has been found that the chair conformation is considerably more stable than the boat form. At room temperature the ratio of chair to boat conformation is about 1000 : 1. This fact has been further established by Hasse (1950).

Examination of the chair form of cyclohexane shows that the twelve hydrogens in the chair conformation may be divided into two categories : six hydrogens (shown by a) point up are called axial hydrogens; the other six hydrogens lie along the equator are called equatorial hydrogens (shown by e). The bonds by which these hydrogens are attached are known as axial and equatorial bonds, respectively. When the chair conformation is converted into boat conformation, each equatorial bond becomes axial and *vice versa*.

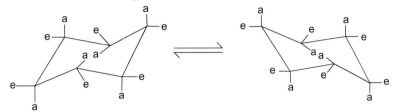

The postulates of Sachse and Mohr were responsible for the synthesis of a number of multi-membered ring compounds. The difficulty encountered in the synthesis of large ring compounds was wrongly ascribed by Baeyer to their instability. In fact, it arises out of the fact that for ring closure to take place it is essential that the two functional groups at the terminal of a bifunctional compound should come sufficiently close so as to react with each other. Thus, if two such groups are situated at the terminal of a long chain, there is less possibility of these two groups coming close enough to react with each other and hence intermolecular condenzation between functional groups of different molecules may result in polymerisation. This difficulty has been recently overcome by high dilution technique.

4.5.4 Relative Stabilities of Cycloalkanes

The relative stability of cycloalkenes is not the same. On the basis of heats of combustion it is found that cyclohexane is the most stable. The relative stability of cyclopropane and cyclobutane is much less than cyclohexane which is due to their cyclic structures and due to their molecules possessing ring strain. This has been experimentally demostrated by examining, their relative heats of combustion. The heats of combution is determined as given below:

4.5.5 Heats of Combustion

The heat of combustion of a compound is the change of enthalpy for complete oxidation of the compound.

In case of hydrocarbons complete oxidation means converting it to CO_2 and H_2O. This is accomplished experimentally and the amount of heat evolved can be measured using a calorimeter. For example in case of methane, the heat of combustion is -803 kJ mol^{-1}.

$$CH_4 + 2O_2 \longrightarrow CO_2 + 2H_2O \qquad \Delta H^\circ = -803 \text{ kJ mol}^{-1}$$

For isomeric hydrocarbons (e.g., butane and isobutane), complete combustion of 1 mol of each will require the same amount of oxygen and will yield the same number of moles of CO_2 and H_2O. Therefore, the measure of the heat of combustion can be used to measure the relative stabilities of the isomers.

$$CH_3CH_2CH_2CH_3 + 6\frac{1}{2}O_2 \longrightarrow 4CO_2 + 5H_2O \qquad \Delta H° = -2877 \text{ kJ mol}^{-1}$$

$$CH_3\underset{\underset{CH_3}{|}}{CH}CH_3 + 6\frac{1}{2}O_2 \longrightarrow 4CO_2 + 5H_2O \qquad \Delta H° = -2868 \text{ kJ mol}^{-1}$$

As seen in the above butane liberates more heat on combustion than isobutane. This implies that butane contains relatively more potential energy. Therefore, isobutane must be more stable.

The cycloalkanes also constitute a homologons series, each member of the series differing from the one immediately preceding of by a constant amount of one $-CH_2-$ group. So, the general equation for the combustion of a cycloalkane will be

$$(CH_2)_n + \frac{3}{2}nO_2 \longrightarrow nCO_2 + nH_2O + \text{heat}$$

Since cycloalkanes are not isomeric, their heat of combustion cannot be directly compared. However, it is possible to calculate the amount of heat evolved per CH_2 group. On the basis of this, the stabilities of cycloalkanes can be compared directly. The results of such investigation are given the table below:

Table 4.3 Heats of comustion and Ring strain of cycloalkanes

Cyclopropane $(CH_2)_n$	n	Heat of combustion (kJ mol⁻¹)	Heat of combustion per CH_2 group (kJ/mol⁻¹)	Ring strain* (kJ mol⁻¹)
Cyclopropane	3	2091	697.0	115
Cyclobutane	4	2744	686.0	109
Cyclopentane	5	3320	664.0	27
Cyclohexane	6	3952	658.7	0
Cycloheptane	7	4637	662.4	27
Cyclooctane	8	5310	663.8	42
Cyclononane	9	5981	664.6	54
Cyclodecane	10	6636	663.6	50
Cyclopentadecane	15	9885	659.0	6
Unbranced alkane		*	658.6	—

* The ring strain of a cycloalkane can be calculated by using the formula:

Ringstrain of a cycloalkane = Heat of combustion of cycloalkane $- [658.7 \times n]$

On the basis of the above results, following generalisations can be made:

(*i*) As seen, cyclopexane has the lowest heat of combustion per CH_2 group (658.7 kJ mol^{-1}). This amount does not differ from that of unbranced alkanes, which have no ring and can have no ring stain. So cycloalkane can serve as a standard for comparison with other cycloalkanes.

(*ii*) Cyclopropane on compustion evolves the greatest amount of heat per CH_2 group. So, molecules of cyclopropane must have the greatest ring strain (115 kJ mol^{-1}) and so cyclopropane molecules must contain greatest amount of potential energy per CH_2 group. What is called the ring strain, it is a form of potential energy that the cyclic molecule contains. The more ring strain a molecule a molecule possess, the more potential energy it has and the less stable it is an compared to its ring homologues.

(*iii*) The combustion of cyclobutane evolves the second largest amount of heat per CH_2 group and so cyclobutane has the second largest amount of ring strain (109 kJ mol^{-1}).

(*iv*) Other cycloalkanes possess to varying degrees, this relative amounts are not large. Cyclopentane and cycloheptane have approximately the same amount of ring strain. Higher alkanes (8, 9 and 10 members) have slightly large amounts of ring strain and then the amount falls off. Finally, a 15 membered ring has only a very slight amount of ring strain.

4.5.6 Chemistry of Cycloalkanes

The cycloalkanes are classified according to the number of carbon atoms present in alanes. The different categories of cycloalkanes are as given below:

Small ring compounds	3 – 4 carbon atoms
Common ring compounds	5 – 7 carbon atoms
Medium ring compounds	8 – 11 carbon atoms
Large ring compounds	> 7 carbon atoms

❑❑❑

5

Chemistry of Small Rings

As already stated small ring cycloalkanes contain 3 and 4 carbon atom, viz, cyclopropanes and cyclobutanes.

5.1 CYCLOPROPANES

We have already discussed the methods of preparation of cyclopropanes and its properties (both physical and chemical) in the earlier chapters of the book.

Given below are some other methods to synthesise cyclopropane and its derivatives.

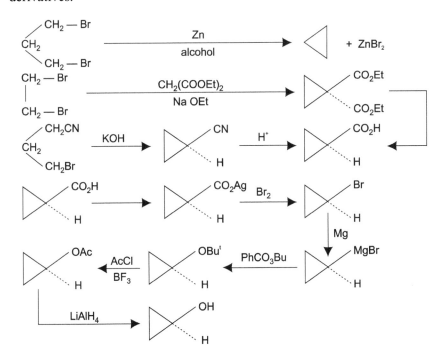

© The Author(s) 2023
V. K. Ahluwalia and R. Aggarwal, *Alicyclic Chemistry*,
https://doi.org/10.1007/978-3-031-36068-8_5

A valuable new method of cyclopropane synthesis involves hydroboration of an allyl halide followed by treatment with alkali.

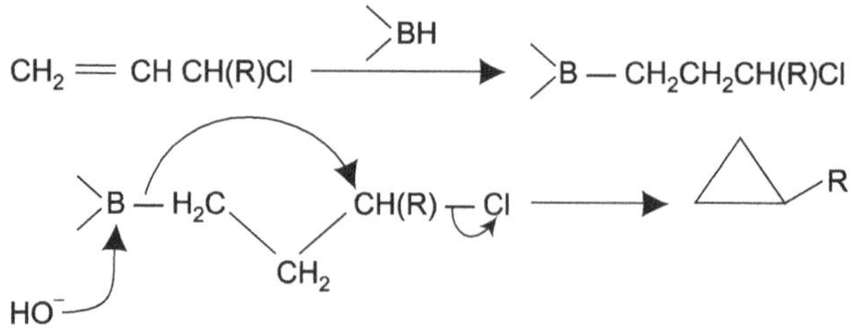

$$CH_2 = CH\ CH(R)Cl \xrightarrow{\quad \overset{\displaystyle >}{}BH \quad} \quad >B - CH_2CH_2CH(R)Cl$$

$$\left[\text{Here, } >BH = \text{9-borabicyclo[3.3.1]nonane}\right]$$

Cyclopropane carboxylic acid can also be prepared by **Favorskii rearrangement** of 2-chlorocyclobutanone.

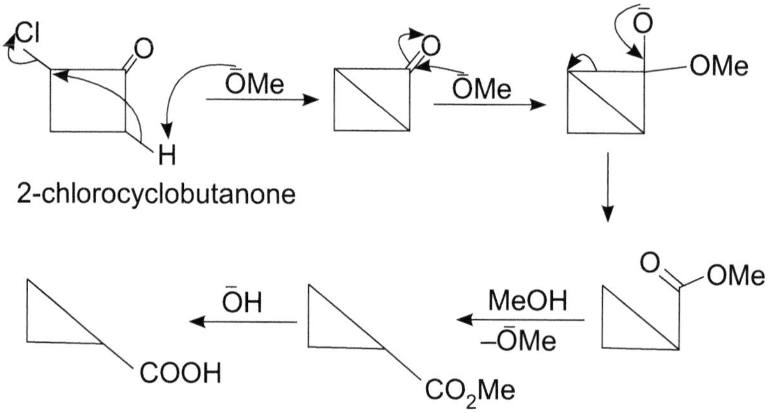

2-chlorocyclobutanone

Cyclopropane
Carboxylic acid

Synthesis of cyclopropanes by carbene addition to an olefin is a very valuable route. The required dichlorocarbene can be generated in situ from chloroform by reacting with t-BuO⁻ or by sonication of chloroform and powdered sodium hydroxide. The last procedure is simple and avoids the use of PTC.

$$t-BuO^- + CHCl_3 \longrightarrow t-BuOH + {}^-CCl_3 \longrightarrow \quad :CCl_2 \quad + \quad Cl^-$$

Dichlorocarbene

$$CHCl_3 + NaOH \xrightarrow{\quad))))\quad} \quad :CHCl_2$$

Dichlorocarbene

In fact dicholro or dibromocarbene adds stereospecifically to olefin.

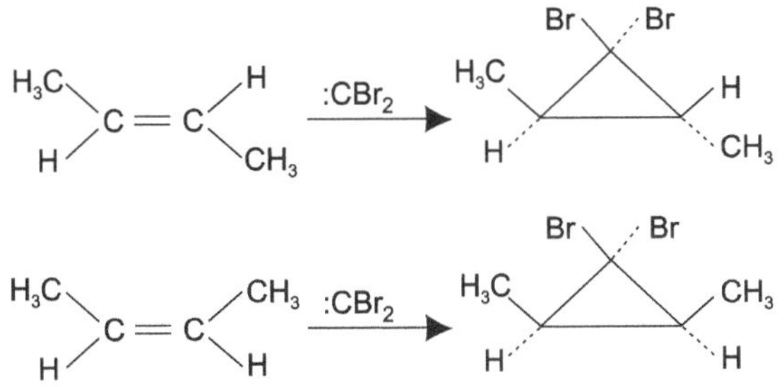

Methylene (:CH$_2$), generated in situ by the photolysis of diazomethane adds mainly stereospecifically. However, in this case insertion into C – H bond constitutes an unwanted side reaction.

$$CH_2N_2 \xrightarrow{\text{h}\nu} \; : CH_2 + N_2$$

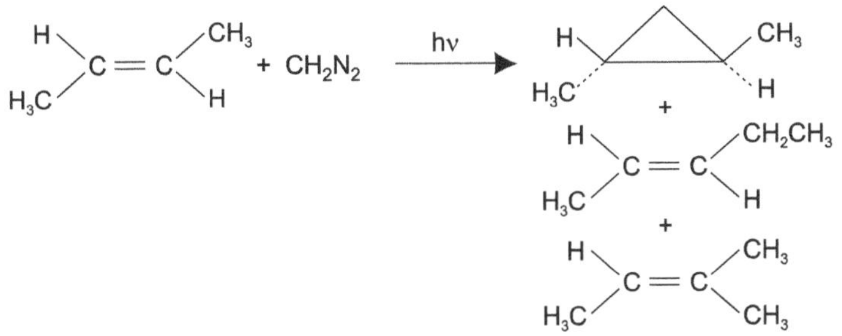

The required diazomethane can also be generated from hydrazine by reaction with potassium hydroxide and choloroform in presence of [18] crown-6

$$H_2NNH_2 \xrightarrow[\substack{[18]\ \text{crown-6}\\ \text{ether}}]{KOH,\ CHCl_3} CH_2 = N^+ = N^-$$

Diazomethane
(yield 48 %)

Alternatively, hydrazine can be reacted with NaOH and CHCl$_3$ in presence of a PTC

$$H_2NNH_2 + CHCl_3 + NaOH \xrightarrow[\substack{\text{ether or}\\ CH_2Cl_2}]{PTC} CH_2N_2 \text{ in ether or } CH_2Cl_2$$

(yield 35 %)

Another convenient source of a methylene reagent, that has the advantage of reacting stereospecifically makes use of CH$_2$I$_2$ with a zinc/copper couple (**Simmons reaction**).

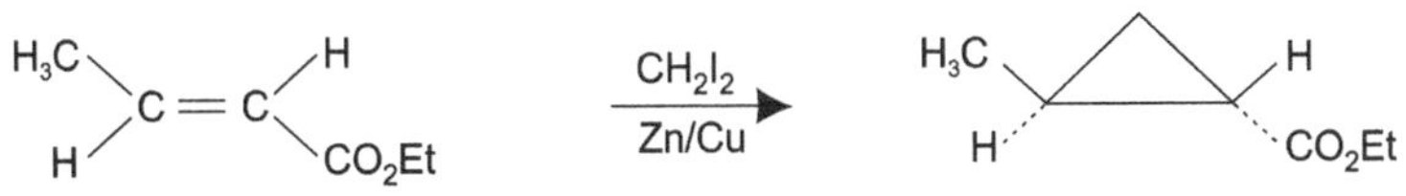

$$CH_3(CH_2)_7 \diagdown_{H} C = C \diagup_{H}^{(CH_2)_7CO_2CH_3} \xrightarrow[\text{Zn/Cu}]{CH_2I_2} CH_3(CH_2)_7 \triangle (CH_2)_7CO_2CH_3$$

Carbene can also be generated by the decomposition of a diazoalkane catalysed by cuprous halide or copper bronze. This gave the products in different ratios, compared to that from photo-decomposition.

$$CH_2 = CHOEt + N_2CHCO_2Et \longrightarrow$$

EtO $\triangle$ H, H···CO$_2$Et	+	EtO $\triangle$ CO$_2$Et, H···H

(a) Cu	$\longrightarrow$	63%	11%
(b) hv	$\longrightarrow$	31%	16%

Addition of diazomethene to an activated alkene followed by thermal or photochemical decomposition of the intermediate pyrazoline derivative gives a cyclopropane derivative. This is also a convenient route for the synthesis of cyclopropanes.

$$EtO_2C \diagdown_{Pr^i} C = C \diagup_{CO_2Et}^{H} \xrightarrow{CH_2N_2} EtO_2C \text{——} H, Pr^i \cdots CO_2Et \xrightarrow[\text{heat or hv}]{-N_2}$$

$$\longrightarrow Et O_2C \triangle H, Pr^i \cdots CO_2Et \quad + \quad EtO_2C \triangle CO_2Et, Pr^i \cdots H$$

A pyrazoline derivative generated from hydrazine and an α, β – unsaturated carbonyl compound also gives a cyclopropane derivative.

$$H \diagdown_{Ph} C = C \diagup_{H}^{COPh} \xrightarrow{N_2H_4} \xrightarrow[\text{Δ or hv}]{-N_2}$$

$$Ph \triangle H, H \cdots Ph \quad + \quad Ph \triangle Ph, H \cdots H$$

An internal carbene reaction may also lead to the formation of cyclopropane derivatives.

$$CH_3 \diagdown_{CH_3}^{CH_3} C - CH_2Cl \xrightarrow{\bar{P}h \overset{+}{N}a} \left[CH_3 \diagdown_{CH_3}^{CH_3} C - CH: \right] \longrightarrow H_3C \diagdown_{H_3C} \triangle \quad 80\%$$

Cyclopropane can also be prepared by the reaction of 1, 3-dibromopropane with Mg in dry ether.

$$Br — CH_2 — CH_2CH_2 — Br \xrightarrow[\text{ether}]{\text{Mg}} Br\ Mg \diagup CH_2 = CH_2 \diagdown Br$$

$$\longrightarrow \quad \underset{\text{CH}_2}{\overset{\text{CH}_2 — CH_2}{\triangle}} \quad + \ Mg\ Br_2$$

5.1.1 Stability of Cyclopropanes

The cyclopropane ring is relatively stable to oxidation, e.g., Carane gives cis-caronic acid in which the cyclopropane ring is intact.

Carane $\xrightarrow[\text{K MnO}_4]{\text{[O]}}$

However, on strong heating, cyclopropanes may get cleaved and rearranged.

Cyclopropane $\xrightarrow{500°C}$ $\xrightarrow{\text{Rearr.}}$ Propene

1,1-Dimethyl cyclopropane $\xrightarrow{500°C}$ $\xrightarrow{\text{Rearr.}}$

Vinyl cyclopropane on heating gives cyclopentene. This rearrangement could be facilitated by allylic stabilisation in a radical intermediate (route A) or 1, 3-concerted migration as in route B.

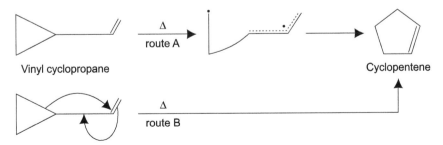

Vinyl cyclopropane $\xrightarrow[\text{route A}]{\Delta}$ Cyclopentene

$\xrightarrow[\text{route B}]{\Delta}$

Substituted cyclopropanes undergo hydrogenolysis. Presence of an electrophilic substituent, e.g., Ph assists hydrogenolysis and also influences the direction of ring fission as shown below:

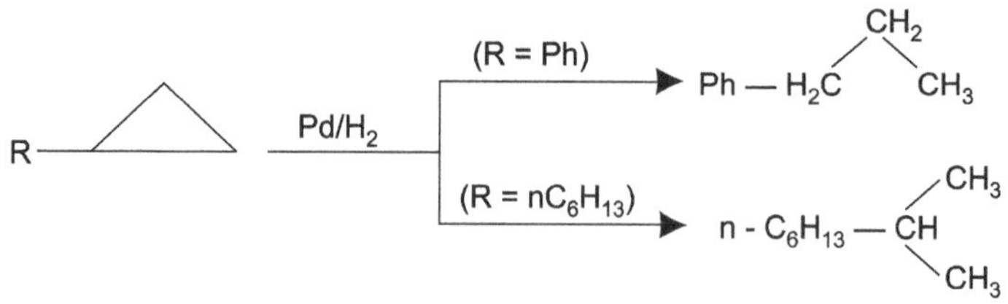

5.1.2 Cis-trans Isomeresm in Disubstituted Cyclopropane

The presence of two substituents in cyclopropane ring gives rise to cis-trans isomenism. As an example 1,2-dimethylcyclopropane can exist in cis and trans forms.

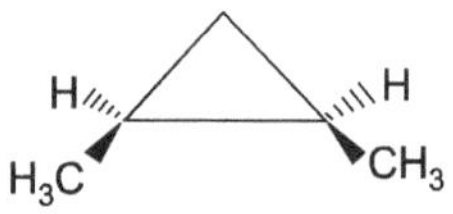

cis-1, 2 - Dimethyl
cyclopropane

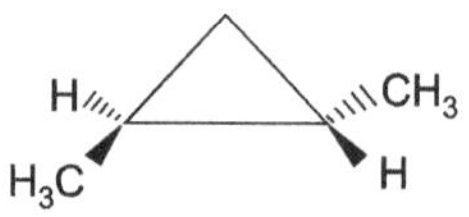

trans-1, 2 - Dimethyl
cyclopropane

The case of cylopropane ring: Banana Bonds

Cyclopropane is an interesting case of a bent C – C covalent bond. It is a planar molecule. The internal bond angle of 60° means considerable deviation from the tetrahedral angle of 109.5°. The carbon atoms cannot have hybridized orbitals oriented at less than 90° to each other. Since there are four bonds on each carbon, the angles will be more than 90°. This means that in cyclopropane, the ring sigma bonds cannot be formed by the bonding atomic orbitals pointing directly at one another. In fact, it has been shown that the orbitals are hybridized in such a way that the orbitals forming the ring have a slightly greater p-character than those forming the normal sigma bonds between carbon atoms in saturated molecules. This explains why the angle between two orbitals of one carbon atom in cyclopropane is 104°, somewhat less than the tetrahedral angle. Thus, it can be understood that the major axes of symmetry of the orbitals forming the C – C bonds in cyclopropane do not point towards one another but outwards from the molecule with the consequence that the C – C bonding MOs are not symmetrical about the internuclear axis but bent giving the appearance of bananas.

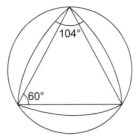

A significant point arising out of higher *p*-character of AOs of the bonded C – C atoms is that there is a small but appreciable degree of pi-delocalisation around the ring, which occurs, unlike in an aromatic system, in the plane of the ring and not above and below the carbon framework. This in-plane pi-delocalisation adds some stability of the ring.

5.1.3 Cyclopropanols

Cyclopropanols are prepared by hydroboration of proprgyl bromide (Brown and Rhodes, 1969)

$$CH \equiv C - CH_2Br \xrightarrow{>BH} (>B)_2CHCH_2CH_2Br \longrightarrow$$

$$\xrightarrow{NaOH} >B-\triangleleft \xrightarrow{H_2O_2} HO\!\!\diagdown\!\!\triangle$$

$$>BH = 9 - borabicyclo [3.3.1] nonane$$

Cyclopropanols readily undergo ring opening by acid or base certalysed reaction.

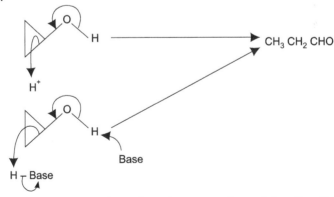

In the acid catalysed reaction, there is protonation of the C – C bond and retention of configuration at the reaction site. In its base induced cleavage, on the other hand, reaction may be regarded as consequent in the development of carbanion character at the reaction site, and proton addition occurs with inversion of configuration. This is illustrated by the following example.

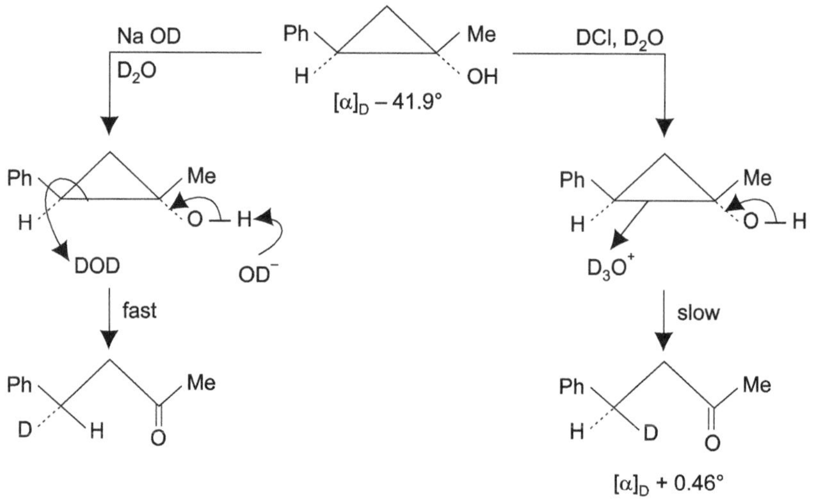

$$[\alpha]_D - 41.9°$$

$$[\alpha]_D + 0.46°$$

Nucleophilic fission of a cyclopropane, carrying a group which stabilises a carbanion, takes place as shown below:

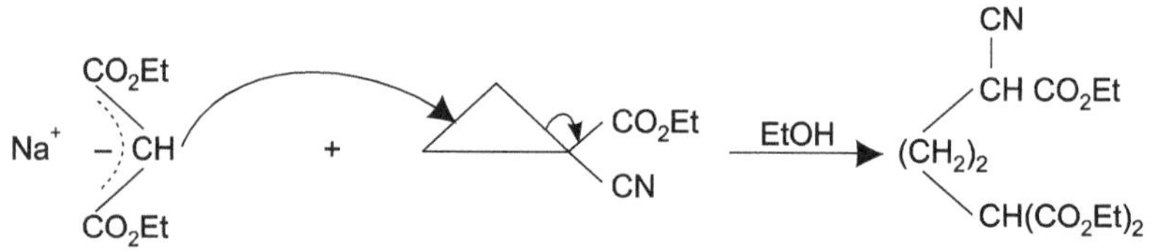

The electrophilic fission of the cyclopropane ring follows the Markownikoff's rule.

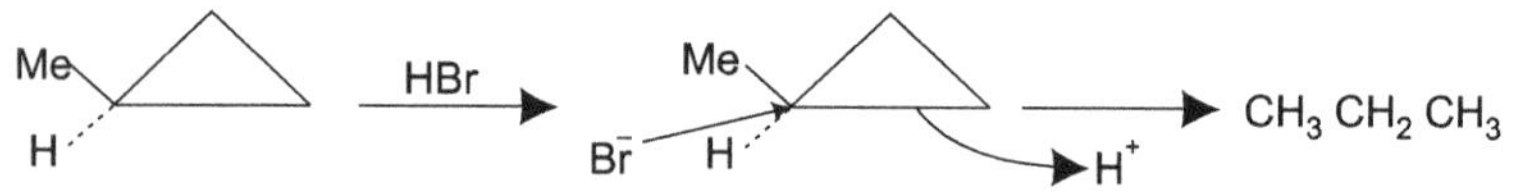

5.1.4 Cyclopropene

In cyclopropene, the olefinic bond is unusually short and the olefinic $v_{CH}[3080 - 3100 \text{ cm}^{-1}]$ is correspondingly high.

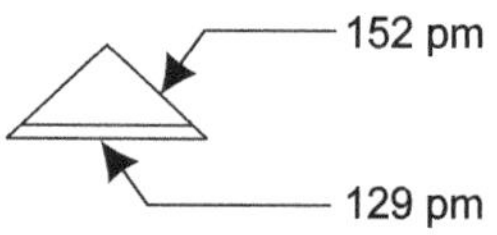

152 pm

129 pm

In view of the above, the olefinic hydrogen displays acidic nature, as is expected of a C − H bound of high s-character.

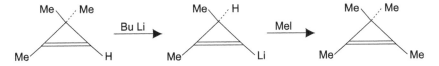

The srained olefinic bond in cyclopropene is also active in the Diels's Alder addition:

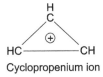

Cyclopropene is obtained by the Hofmann elimination as follows:

Cyclopropene derivatives can also be obtained by making use of the principle of internal carbene addition.

5.1.4.1 *Cyclopropenium ion*

The cyclopropenium ion represents the simplest aromiatic ring structure of $(4n + 2)\pi$ electrons, $n = 0$, in terms of Huckel's definition (Breslow, 1970).

Cyclopropenium ion

A number of cyclopropenium salts have been synthesised by the addition of a suitable carbene to an acetylene.

Thus, phenylchlorocarbene, generated from benzylidene chloride, adds on to diphenylacetylene to give triphenyl-cyclopropene, which is converted into triphenyl cyclopropenium chloride as shown below:

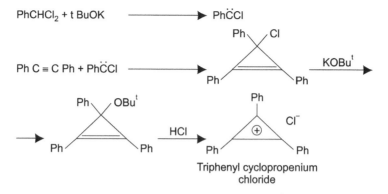

In a similar way, ethoxycarbonyl carbne, generated by the decomposition of ethyl diazoacetate to oct-4-yne, gives ethoxycarbonyl di-n-propl cyclopropene, which is then converted into the cyclopropenium salt.

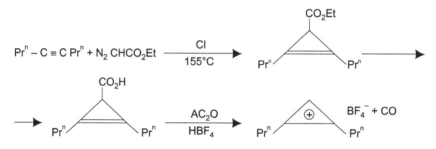

The mechanism of the last step is of interest and is given below:

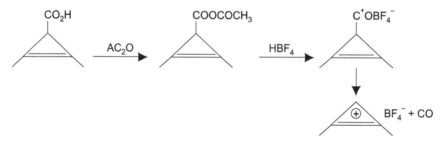

Some salts of unsubstituted cyclopropenium ion have been obtained from chlorocyclopropene.

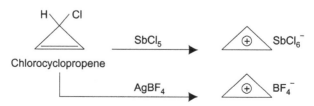

The starting chlorocyclopropene is obtained as follows:

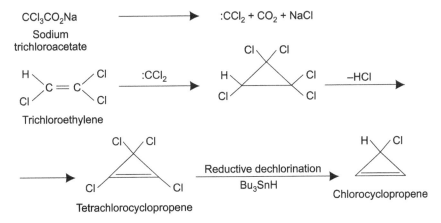

Trichloroethylene

Tetrachlorocyclopropene

Reductive dechlorination
Bu₃SnH

Chlorocyclopropene

5.1.4.2 *Cyclopropenyl ion*

The cyclopropenyl ion (represented below) difers from cyclopropenium ion in that the latter conforms to Huckel's $4n + 2$ π-electrons criteria of aromaticity for $n = 0$, while the anion represents a $4n\pi$-electrons system ($n = 1$), which the anion represents a $4n$, π-electrons system ($n = 1$), which is regarded as antiaromatic and relatively destabilised. Due to this reason, the CH_2 group of a cyclopropene is a weaker carbon acid than the CH_2 group of a similar cyclopropane.

(Cyclopropenyl anion)   Cyclopropenium cation

5.2 CYCLOBUTANES

Cyclobutanes can be prepared as given below:

(a) By the reaction of α, ω-dibromoalkane with disodiomalonic ester. (for details see page 7).

(b) By Hofmann exchaustive methylation method (Willstater, 1907).

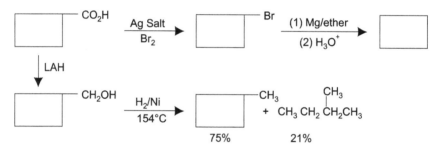

The cyclobutane carboxylic acid obtained can also be converted into cyclobutane as follows:

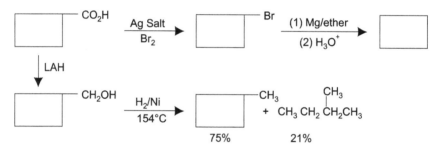

75%   21%

(c) Some cyclobutane derivatives may be prepared by the cyclo addition reaction as follows:

The above reaction proceeds satisfactorily when tetrafluoroethylene is the dienophile.

(d) A valuable method involves dimerisation of highly reactive olefins. Thus, dimethyl ketene dimerises spontaneously to give cyclobutane derivative.

In the above dimerisation, the orientiation is exceptional to give 1, 4-substituted products. However, in most of the cases dimerisation gives mainly 1, 2-disubstituted cyclobutanes. Some examples are given below:

(e) Photo-addition of α, β-unsaturated ketones with ethylenes.

5.2.1 Properties

(a) *Action of heat*

In general, cyclobutane is resistant to thermolysis, though it may get decomposed into ethylene at about 500°C.

Presence of vinyl group in cyclobutane assists the thermolysis.

Vinyl cyclobutane

In a similar way, pyrolysis of α-pinene (which may be considered to have the vinyl cyclobutane system takes place easily; this is an industrially important process.

α-Pinene

Cyclobutane can also be obtained from cyclobutylbromide by converting it into cyclobutyl magnesium bromide followed by treatment with water.

(b) *Displacement reactions*

In displacement reactions, cyclobutyl derivatives usually undergo rearrangements.

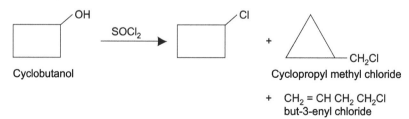

Cyclobutanol

Cyclopropyl methyl chloride

+ $CH_2 = CH\ CH_2\ CH_2Cl$
but-3-enyl chloride

2-Bromocyclobutanone undergoes **Favorskii rearrangement** to cyclopropane carboxylic acid

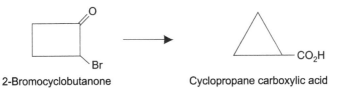

2-Bromocyclobutanone   Cyclopropane carboxylic acid

(c) *Oxidation*

Cyclobutane derivatives are stable to oxidative degradation as is illustrated by oxidative degradation of α-pinene to pinonic acid.

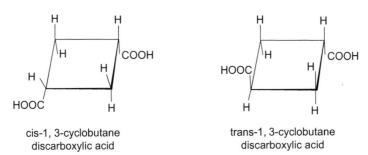

α-Pinene   Pinonic acid

Cis-trans isomerism in disubstituted cyclobutane

In cyclobutane cis-trans isomerosm is possible in case the ring contains two substituents. As as example 1, 3-cyclobutane dicarboxylic acid can have cis and trans forms

cis-1, 3-cyclobutane
discarboxylic acid

trans-1, 3-cyclobutane
discarboxylic acid

5.2.1.1 *Cyclobutanone*

It is obtained from pentaerythritol bromide by reacting with zinc or from Ketene with diazomethane

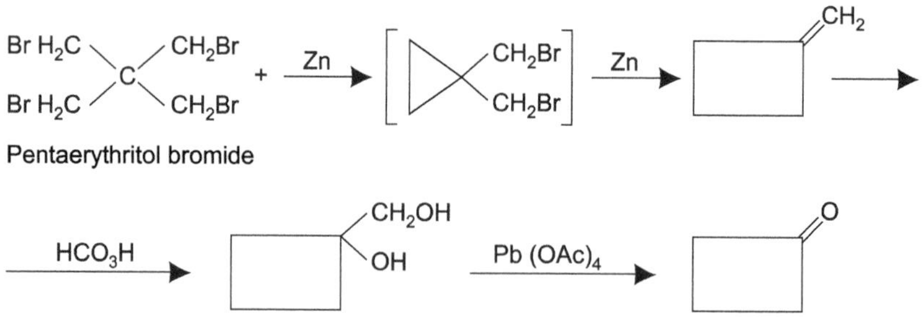

Pentaerythritol bromide

cyclobutanone

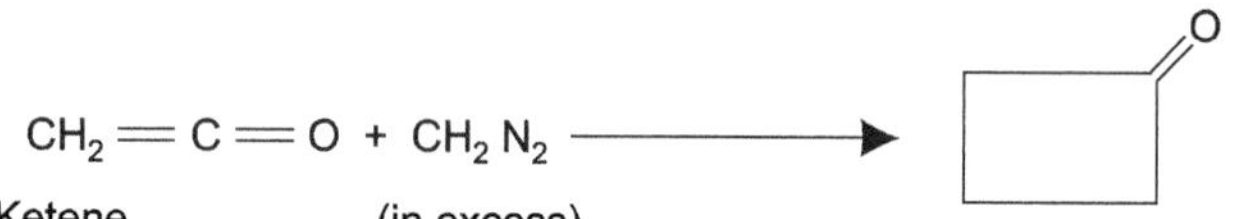

$CH_2 = C = O + CH_2 N_2 \longrightarrow$

Ketene   (in excess)

cyclobutanone

Cyclobutanone on reduction with lithium aluminium hydride gives cyclobutanol.

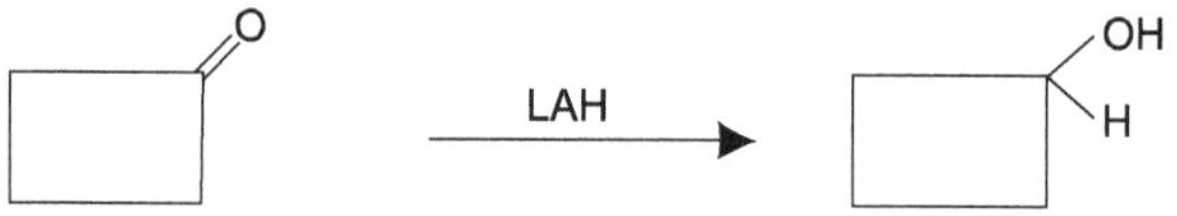

Cyclobutanone   Cyclobutanol

5.2.1.2 Cyclobutene

It is obtained by the Hofmann or cope elimination reactions.

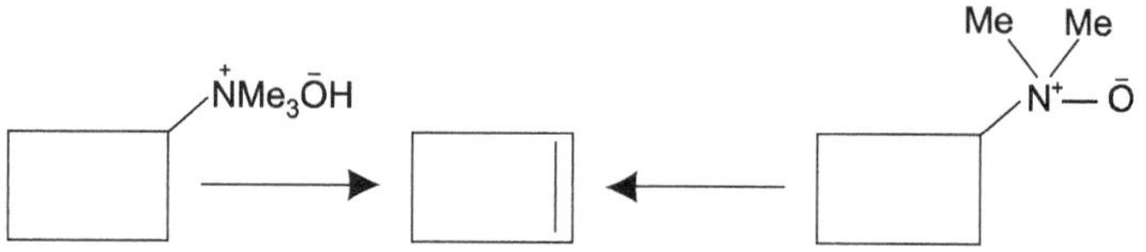

Cyclobutanone is readily converted into buta-1, 3-diene.

Cyclobutene   Butadiene

□□□

Chemistry of Common Ring Compounds

The compounds that belong to this category are cyclopentane, cyclohexane and cycloheptane and its derivatives.

6.1 CYCLOPENTANES

As already stated, cyclopentane and its derivatives are obtained from diethylmalonate and 1, 4-dibromobutane in presence of NaOEt to give cyclopentane carboxylic acid. Thry are also obtained by heating calcium, barium or thorium salts of adipic acid. Besodls these methods, cyclopentane and its derivatives are also obtained by the following methods.

(a) From adipic acid and diethyl adipate. Various steps involved are given below:

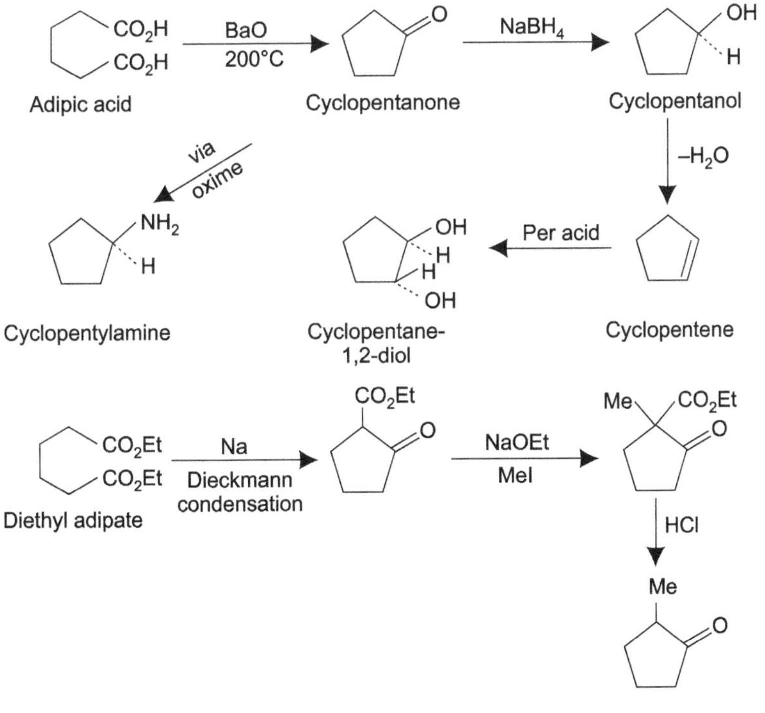

© The Author(s) 2023
V. K. Ahluwalia and R. Aggarwal, *Alicyclic Chemistry*,
https://doi.org/10.1007/978-3-031-36068-8_6

(b) *From diethyl malonate and 1, 4-dibromobutane.*

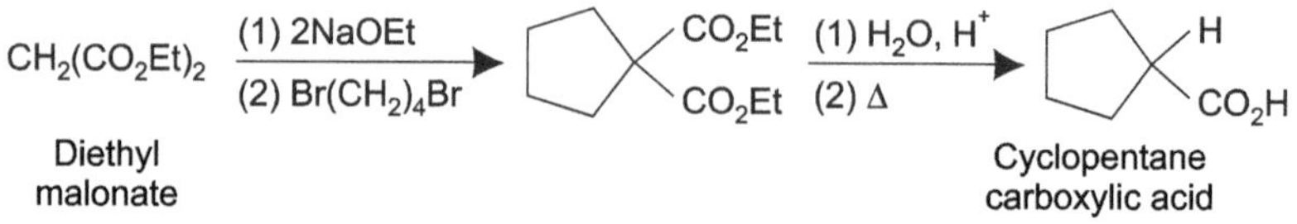

CH$_2$(CO$_2$Et)$_2$ $\xrightarrow[\text{(2) Br(CH}_2\text{)}_4\text{Br}]{\text{(1) 2NaOEt}}$ [cyclopentane with CO$_2$Et, CO$_2$Et] $\xrightarrow[\text{(2) } \Delta]{\text{(1) H}_2\text{O, H}^+}$ [cyclopentane with H, CO$_2$H]

Diethyl
malonate

Cyclopentane
carboxylic acid

(c) From diethyl oxalate and diethyl glutarate by intermolecular condensation.

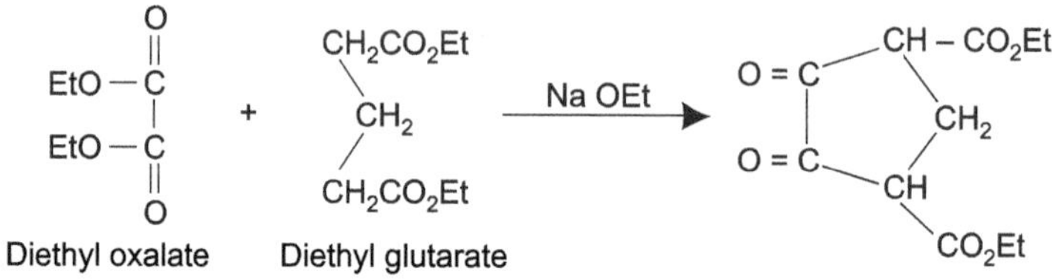

Diethyl oxalate   Diethyl glutarate

(d) *From cyclohexanol*

Cyclohexanol on oxidation gives adipic acid, which on heating with baryta gives cyclopentanone. The method involves ring contraction.

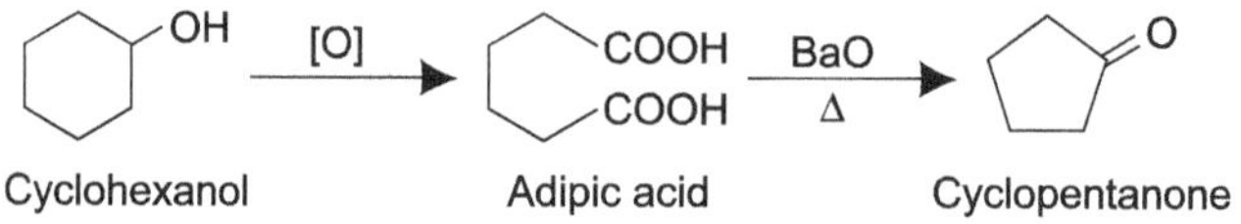

Cyclohexanol   Adipic acid   Cyclopentanone

(e) *Favorskii rearrangement*

Base catalysed rearrangement of α-halo cyclic ketones (known as Favorskii rearrangement) also involves ring contration. Thus, the reaction of 2-bromocyclohexanone (an α-halocyclic ketone) on treatment with sodium methoxide gives methyl cyclopentane carboxylate.

2-Bromo cyclohexanone

Methyl cyclopentane
carboxylate

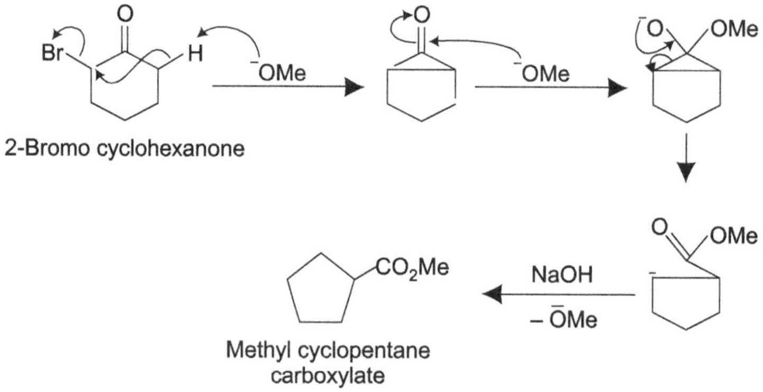

(f) *Wagner Meerwein rearrangemnet*

Cyclobutyl carbinol on treatment with hydrogen bromide undergoes Wagner Meerwein rearrangement to give bromocyclopentane

Cyclobutylcarbinol   Bromocyclopentane

cis-trans isomerism in distributed cyclopentane

In case cyclopentane contains two substituents, cis-trans isomerism is possible. As an example 1, 2-dimethylcyclopentane can exist in cis and trans forms.

cis- 1, 2-Dimethyl
cyclopentane

trans- 1, 2-Dimethyl
cyclopentane

6.2 CYCLOHEXANES

These are prepared by the following methods:

(a) *Catalytic reduction of aromatic substances.*

HO_2C — ⟨ ⟩ — OH $\xrightarrow{H_2/Ru}$ HO_2C — ⟨ ⟩ — OH

If hydroxyl group is present in the aromatic compound, ruthenium is preferred as catalyst, since platinum may cause hydrogenolysios of the hydroxyl group.

Hydrogenation of resorcinol in alkaline medium gives cyclohexane-1, 3-dione or dihydro resorcinol directly.

OH
$\xrightarrow[\text{NaOH}]{H_2/\text{Ni or Rh}}$

Resorcinol   Dihydro resorcinol

Reduction of aromatic ring using sodium or lithium in liquid ammonia with an alcohol, commonly known as **Birch reduction**, gives 1,4-dihydro benzene derivative (if the substituent R is an electron withdrawing group, e.g., COOH)

or 2,5-addition product (if the substituent R is an electron donating group, e.g., OMe, Me, etc.)

1,4-addition product
(R = electron withdrawing group)

2,5 addition product
(R = electron donating group)

(b) Cyclisation methods

(i) Intramolecular cyclisation (intramolecular Wurtz reaction) of 1,6-dibromohexane with Zn gives cyclohexane (see page 8).

(ii) The reation of diethyl malonate with α, ω-dibromo-pentane or its derivative in the presence of sodium ethoxide gives substituted cyclohexane carboxylic acid.

4, 4-Dimethylcyclohexane
carboxylic acid

(iii) Dieckman cyclisation of ethyl succinate in the presence of sodium ethoxide gives cyclohexane-2, 5-dione-1, 4-dicarboxylic ester (see page 10).

(iv) Cyclisation may be brought about by generating an electrophilic centre in 1, 6-addition to an olefinic bond. This is illustrated by the cyclisation of (–) linalool to give (+)-α-terpinyl acetate (along with some other products).

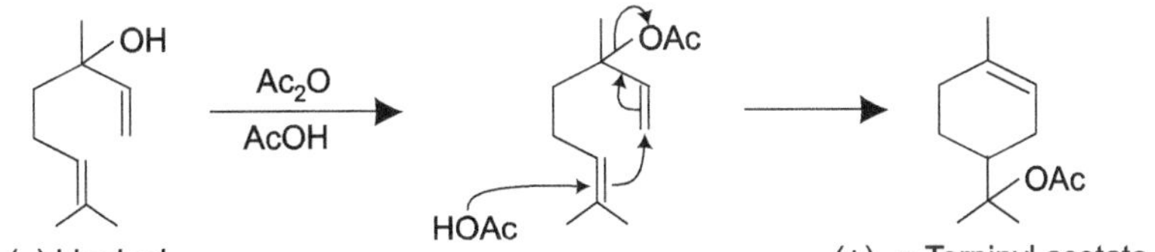

(−) Linalool   (+) -α-Terpinyl acetate

(v) Michael addition of mesityl oxide to sodio malonic ester followed by Claisen condensation gives dimidone.

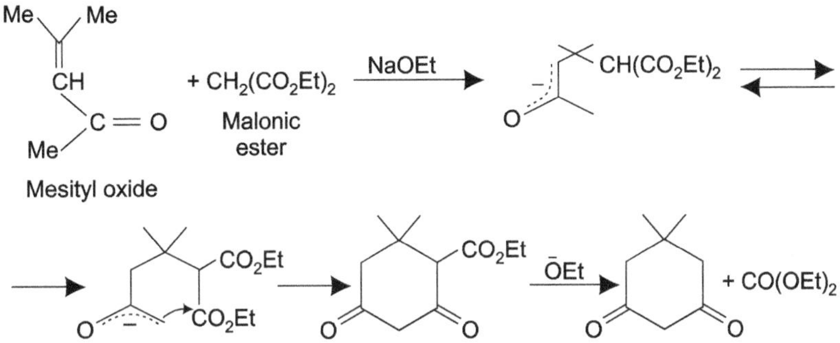

Mesityl oxide

Dimidone

(vi) **Diels'-Alder synthesis.** It is a valuable route for the synthesis of six-numbered ring. Two well known examples are given below:

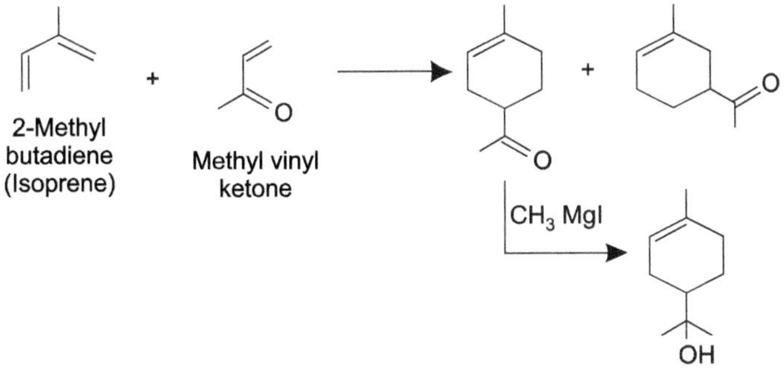

2-Methyl butadiene (Isoprene)

Methyl vinyl ketone

(±)-α-Terpineol

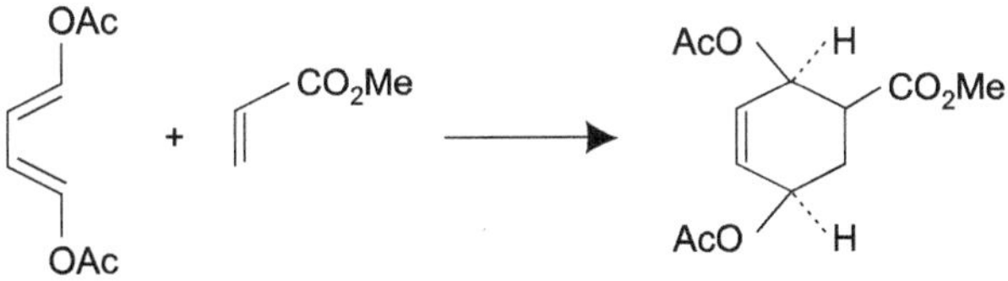

(*vii*) Synthesis of cis- and trans- cyclohexane-1, 2-diol is carried out as given below, starting from cyclohexanol.

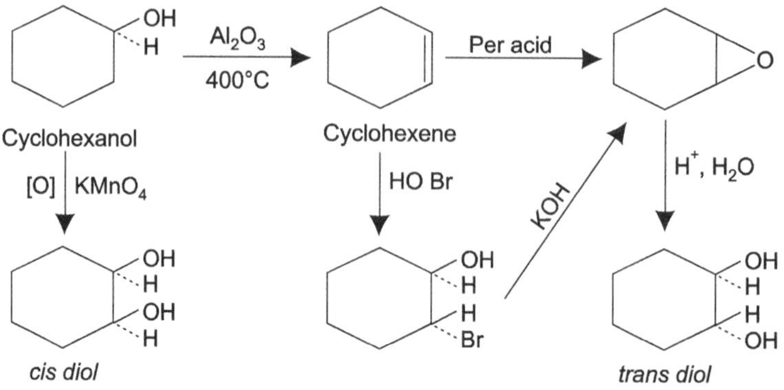

Cyclohexanol   Cyclohexene

cis diol   trans diol

6.3 CYCLOHEPTANE

Derivatives of cycloheptane are obtained by the following methods:

(a) *Ring homologisation*

The reation of cyclohexanone with diazomethane gives cycloheptanone in good yield. A small amount of an epoxide is obtained as a by product.

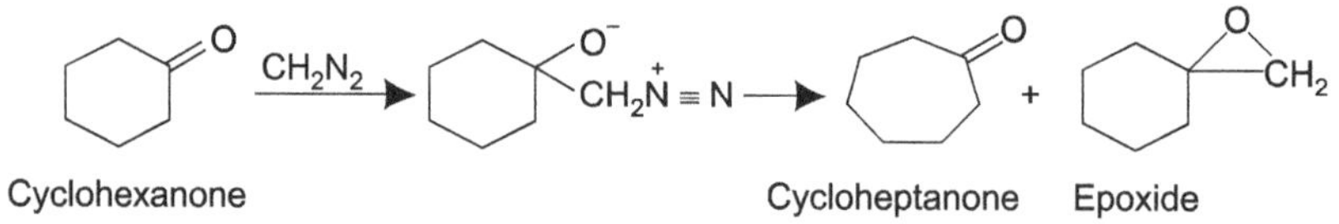

Cyclohexanone   Cycloheptanone   Epoxide

(b) *Demjanov- Tiffeneau ring expansion*

This is based on deamination and pinacolic rearrangement.

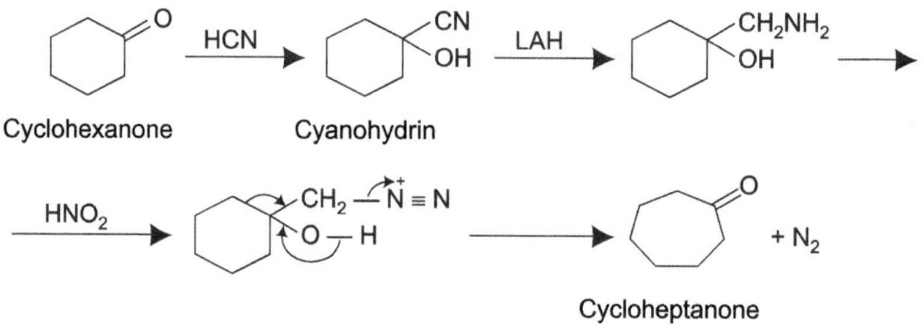

Cyclohexanone   Cyanohydrin

Cycloheptanone

Cycloheptanone can also be obtained from cyclohexanone by treatment with diazomethane in alcoholic solvents as given below.

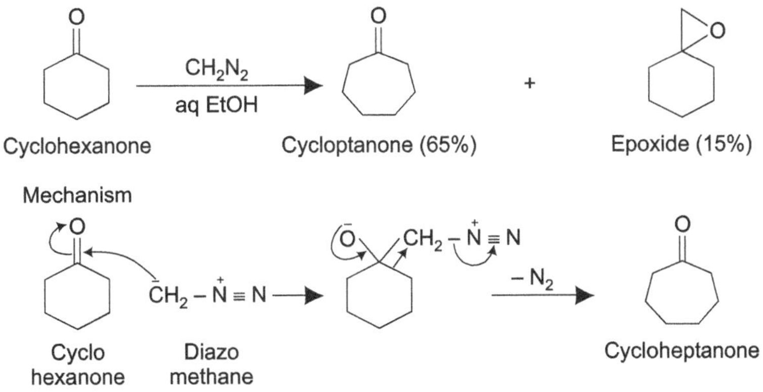

Cyclohexanone   Cycloptanone (65%)   Epoxide (15%)

Mechanism

Cyclo   Diazo   Cycloheptanone
hexanone   methane

6.4 CYCLOHEPTATRIENE

(a) It is obtained by ring homologisation of benzene with diazomethane.

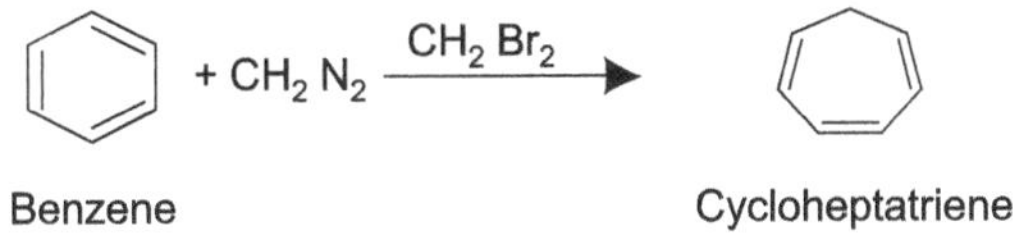

Benzene   Cycloheptatriene

Ethyl diazo
Benzene   acetic ester   Ethyl cycloheptatriene
                                 carboxylate

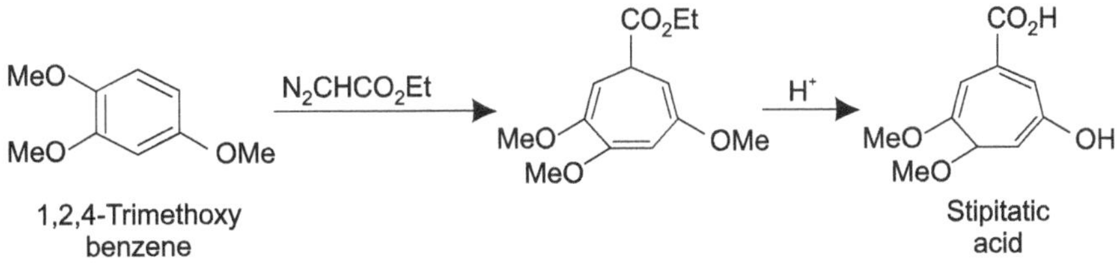

1,2,4-Trimethoxy   Stipitatic
benzene   acid

(b) Cycloheptatriene can also by synthesised by a convenient method involving ring expansion by Wagner-Meerwein rearrangement.

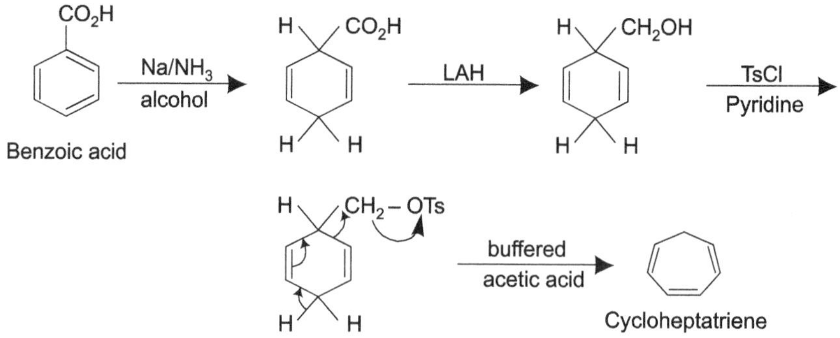

Benzoic acid

Cycloheptatriene

This method can also be used for the synthesis of tropolone.

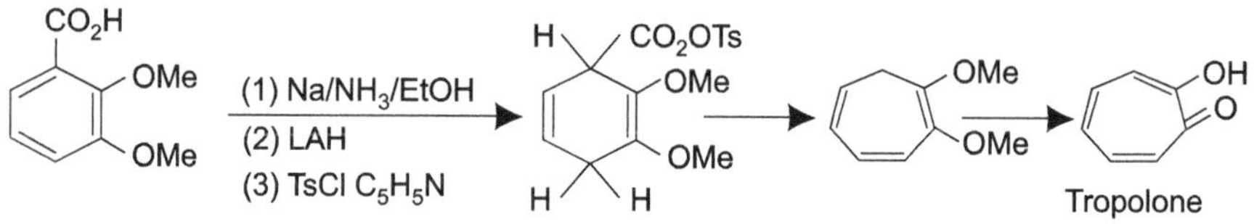

Tropolone

The cycloheptatrience molecule is non-planar with rapid inversion between the following two equivalent structures A and B.

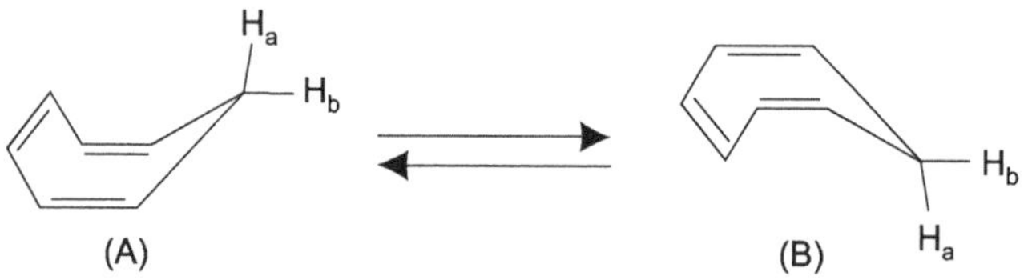

(A)   (B)

The inversion is slow at low temperature; this is seen by the NMR proton signals for Ha and Hb (which are resolved) at low temperature. From its temperature dependence of the spectrum, the activation energy for its inversion A ⇌ B is found to be 26.5 kJ/mole. Cycloheptatriene also exhibits thermal isomerisation due to 1, 5-hydrogen transfer across the molecule. Thus, 7-deuterocyclo heptatriene undergoes isomerisation, the process can be followed by the change in the NMR signal for allylic and vinylic protons.

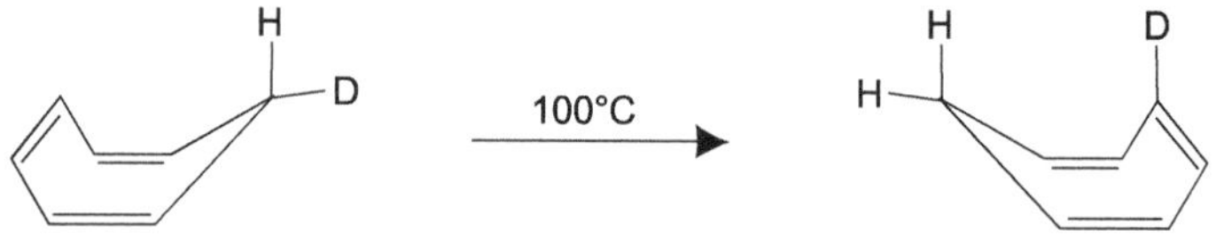

100°C

7-Deutero cycloheptatriene

The type of rearrangement described above is an example of sigmatropic reactions designed by Woodward and Hoffmann; in this, a sigma bond changes its position in an adjacent π-electron system.

❏❏❏

7

Chemistry of Medium Sized and Larger Rings

Medium sized ring include cyclooctane and large size rings include civatone and Muscone

7.1 CYCLOOCTANE AND ITS DERIVATIVES

These are prepared as given below:

(*a*) Dimerisation of butadiene

Butadiene on dimersation gives the following products:

| Butadiene | 4-Vinylcyclohexene (I) | 1,2-divinylcyclo butane (II) | cis-1,2-divinylcyclo butane (III) |

Thermal dimerisation gives 4-vinylcyclohexene (I) as the major product. However, photochemical dimerisation in the presence of a photosensitier gives divinylcyclobutane (II) (arising by a 2 + 2 cycloaddition process) as the major product. The cis-1, 2-divinylcyclobutane (III) can be easily thermally rearranged to cycloocta-1, 5-diene (IV).

| III cis 1,2-divinylcyclobutane | IV Cycloocta-1,5-diene |

Small amount of IV is also obtained in the thermal dimeristaion of butadiene.

Polymerisation of butadiene can also be achieved via a Ni^o–butadiene complex, which may be formed *in situ* by ligand displacement from, $Ni(CH_2CHCN)_2$ or from a Ni(II) salt like acetate or acetylacetonate

© The Author(s) 2023
V. K. Ahluwalia and R. Aggarwal, *Alicyclic Chemistry*,
https://doi.org/10.1007/978-3-031-36068-8_7

reduced in situ by aluminium alkyl or a metal hydride. The intermediate (X) formed reacts with triphenylphosphine to give cycloocta-1, 5-diene.

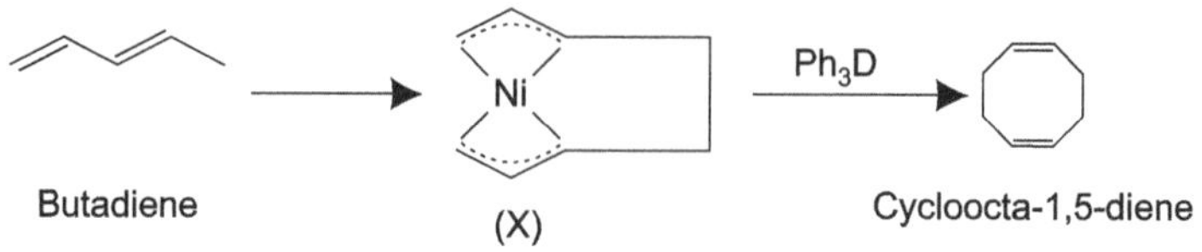

Butadiene   (X)   Cycloocta-1,5-diene

The intermediate, (X), may react further with butadiene to give *cis*, *trans*, *trans*-cyclododecatriene.

cis, trans,
trans-cyclododecatriene

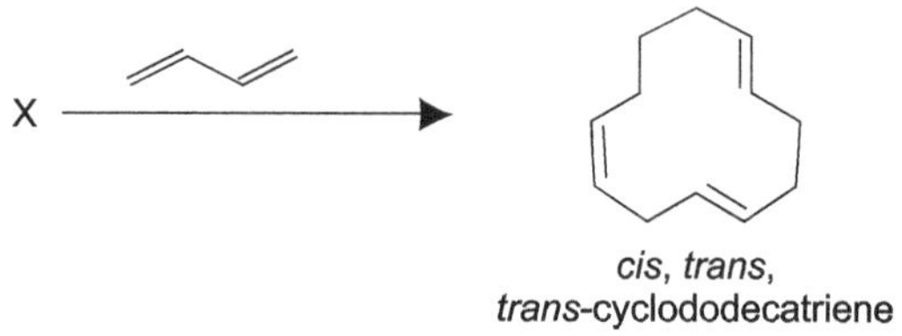

Polymerisation of acetylene can also be catalysed by suitable Ni (II) complexes, e.g., the cyanide, acetylacetonate or salicaldimine derivative, to give cyclo-octatetraene in high yields. The polymerisation proceeds through an intermediate nickel complex (Y) bearing four co-ordinated acetylenes.

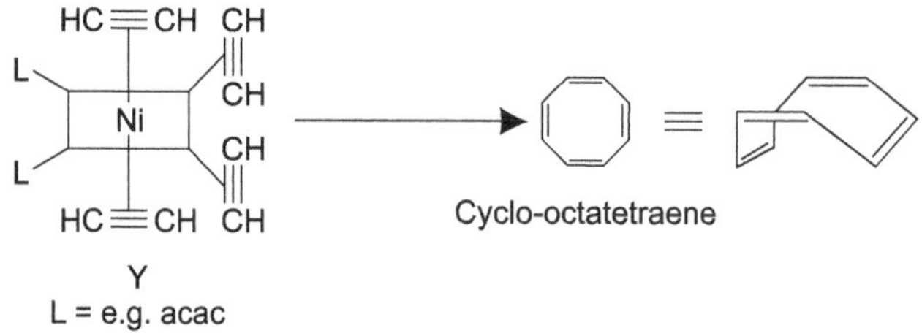

Cyclo-octatetraene

Y
L = e.g. acac

(*b*) Photoaddition of benzene to a suitably substituted acetylene gives cyclo-octatetraene derivative.

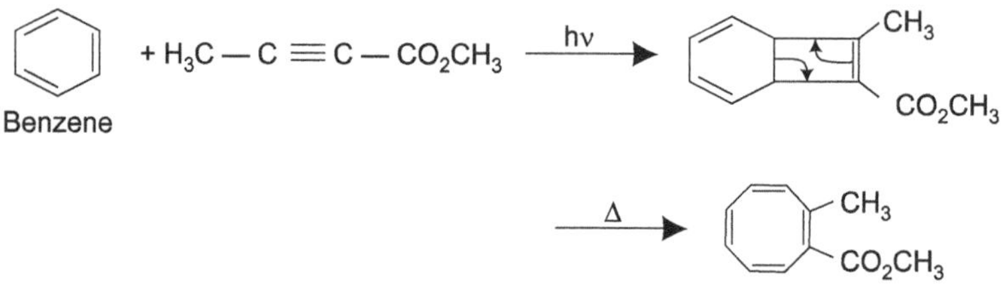

Benzene

Dimerisation of cyclobutene-1, 2-dicarboxylic ester followed by thermal rearrangement of tricyclo-octane derivative gives cyclo-octadiene derivative.

Methyl cyclobutene-
1,2-dicarboxylic acid ester

Cyclo-octadiene
tetracarboxylic ester

(c) Annelation of ethyl cyclohexanone-2-carboxylic ester with acraldehyde followed by ring scission gives ethyl cyclooctene-1, 5-dicarboxylate.

Ethyl cyclohexane
2-carboxylate

Acrolein

Ethyl cyclooctene-
1,5-dicarboxylate

The final step of ring scission is dependent on β-keto-ester fission:

A different approach makes use of an enamine.

Ethyl cyclooctene-5-carboxylate

The final step of ring scission depends on the fragmentation process.

(d) Large ring compounds can be obtained by homologation of cyclo-octene.

Cydooctene

allene   1,3-cyclononadiene

(e) Ring expansion by two carbon atoms is possible by cycloaddition between a cycloalkenamine and an acetylenic ester via rearrangement of the intermediate cyclobutene.

Cycloalkenamine   Acetylenic ester

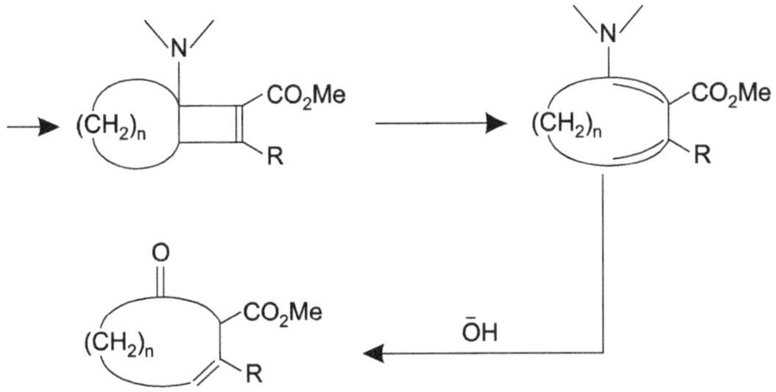

The cyclobutene-diene rearrangement is characteristic of cyclobutenes. Similarly, use of cycloheptenamine and methyl propiolate gives cyclononanone.

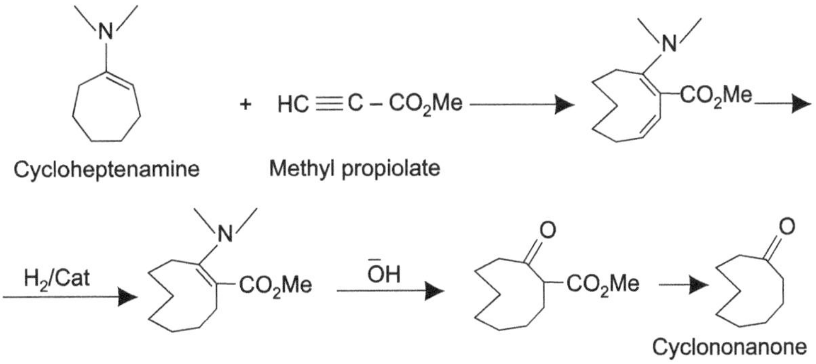

Cycloheptenamine   Methyl propiolate

Cyclononanone

(f) Acyloin synthesis

This is a very convenient method for the synthesis of medium and large ring compounds.

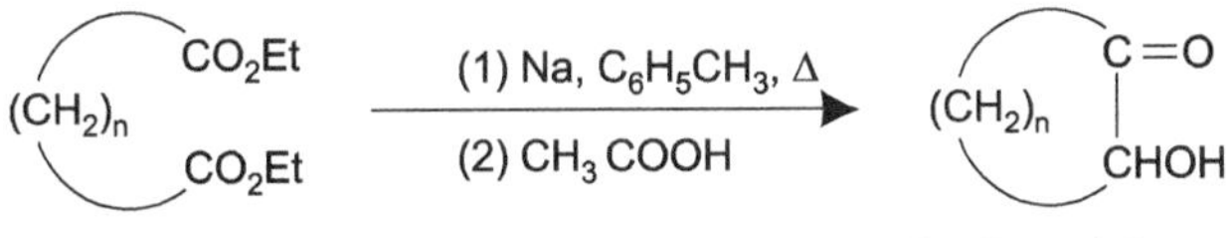

Cyclic acyloin

(for more details and the mechanism of the reaction, see page 13).

The yield in acyloin condensation is 40-50% in the synthesis of eight- or nine-membered rings; with larger rings, for example, C_{21}-acyloin, yields of 90% are possible.

Acyloin synthesis is a convenient source of other derivatives as given below:

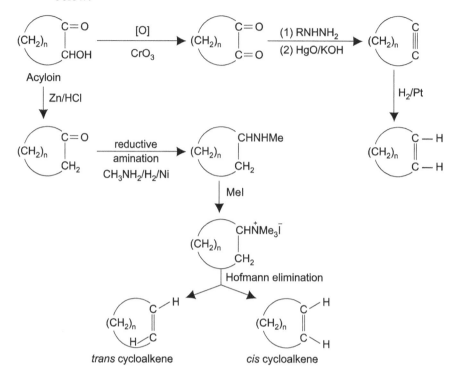

7.2 CIVETONE

It is the active principle of civet from the scent gland of the African civet cat Civetone (and also muscone, isolated from a scent gland of the Tibetan musk deer) is used in preparation of perfumes. Although civetone (and also muscone) does not have pleasant odour, it has the property of markedly enhancing and increasing the persistence of the flower essences.

The crude civetone obtained from commercial civet was purified by hydrolysis with alcoholic potassium hydroxide followed by fractional distillation of the unsaponifiable neutral material under reduced pressure, and treatment of the distillate (b.p. range 140-180°/3 min) with semicarbazide hydrochloride in presence of acetate ion. The semicarbazone of civetone thus obtained on decomposition by boiling with oxalic acid followed by distillation under reduced pressure gave crystalline civetone, m.p. 31°C.

The structure of civetone was elucidated by Ruzicka as given below:

(*i*) The molecular formula of civetone was found to be $C_{17}H_{30}O$, on the basis of elemental analysis and molecular weight determination.

(ii) The oxygen is present as an oxo group (aldehyde or ketone) indicated by the formation of an oxime and it is present as a keto group is shown by (vi) below.

(iii) It contains an olefinic bond as indicated by decolourisation of bromine on addition of bromine.

(iv) It contains a double bond, since on catalytic reduction, civetone absorbs one molecule of hydrogen to give dihydrocivetone ($C_{17}H_{32}O$).

(v) The fully saturated hydrocarbon obtained by reduction of keto group (to CH_2 group) and the reduction of the olefinic bond has the molecular formula $C_{17}H_{34}$. This formula corresponds to a monocyclic hydrocarbon.

(vi) Dihydrocivetone condenses will benzaldehyde indicating the presence to a — CH_2 — CO group. Also, the oxidation of dihydrocivetone with chromic acid gives a dicarboxylic acid ($C_{17}H_{32}O_4$), without loss of carbon, the oxo group is present as a keto group, which must be present in the ring. Further more, the same dicarboxylic acid is obtained by Clemenson's reduction of civetone to civetene (in which the double bond remains intact), followed by ozonolysis. So, the double bond is also present in the ring.

The dicarboxylic acid ($C_{17}H_{32}O_4$) obtained above by both the routes was shown to be identical with pentadecane 1,15- dicarboxylic acid (by comparison with a synthetic specimen).

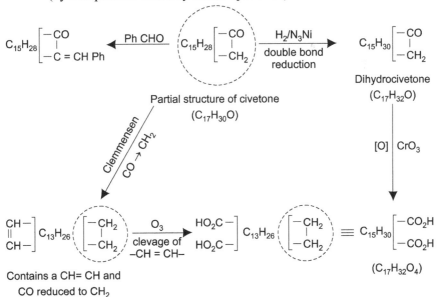

Partial structure of civetone
($C_{17}H_{30}O$)

Dihydrocivetone
($C_{17}H_{32}O$)

($C_{17}H_{32}O_4$)

Contains a CH= CH and
CO reduced to CH_2

On the basis of the above studies, it was found that civetone contains a seventeen-membered ring along with a keto group and a double bond. The above facts may be formulated on the basis of the partial structure of civetone given on page 58.

(vii) The relative positions of the keto group and the double bond was established in the following way :

Oxidation of civetone with cold potassium permangnate gave a ketodicarboxylic acid, $C_{17}H_{30}O_5$. Since the keto group is intact, it is further proved that the double bond is present in the ring. Oxidation of ketocarboxylic acid with sodium hypobromite results in a mixture of succinic, pimelic, suberic and azelic acids. From the formation of azelic acid $HO_2C(CH_2)_7\,CO_2H$, it was inferred that the double bond in civetone is separated from the keto group by at least seven methylene groups. Also, the absence of higher acids, indicates a symmetrical structure. All these facts suggest the structure (I) for civetone.

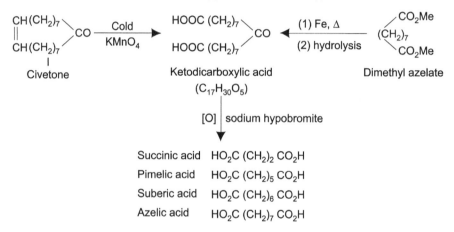

Succinic acid $HO_2C\,(CH_2)_2\,CO_2H$
Pimelic acid $HO_2C\,(CH_2)_5\,CO_2H$
Suberic acid $HO_2C\,(CH_2)_6\,CO_2H$
Azelic acid $HO_2C\,(CH_2)_7\,CO_2H$

The structure (I) for civetone is also supported by the synthesis of ketocarboxylic acid from methyl ester of azelic acid. On the basis of the above discussion Ruzicka inferred that civetone is 9-cycloheptadecenone as shown in structure (I).

(viii) The structure (I) of civetone was also confirmed by oxidation of dihydrocivetone by chromic acid to heptadecanedioic acid.

(ix) Further evidence for the confirmation of the structure of civetone was obtained by the following series of transformations.

$$O=C \underset{(CH_2)_7}{\overset{(CH_2)_7}{\diagup}} \begin{matrix} CH \\ \| \\ CH \end{matrix} \xrightarrow[\text{Zn(Hg)/HCl}]{\text{Clemmensen redn.}} CH_2 \underset{(CH_2)_7}{\overset{(CH_2)_7}{\diagup}} \begin{matrix} CH \\ \| \\ CH \end{matrix}$$

(I)
Civetone

Civetene

$\downarrow$ H$_L$/Pt

Dihydrocivetol

$$\xrightarrow[-H_2O]{KHSO_4,\ 180°C}$$

Civetene

The above transformations show that the symmetry of the civetone ring is such that civetene (cycloheptadecane) produced by the Clemmensen reduction of civetone is identical to civetene obtained by reduction of civetone to dihydrocivetol and dehydration.

(x) Final confirmation of the structure of civetone is obtained by its synthesis. A number of procedures have been carried out. The one given by Stoll et al., (1948) is discussed below. This procedure involves acyloin synthesis, protection of the keto group and formation of cis- and trans- isomers.

$$OC \underset{(CH_2)_7\ CO_2Me}{\overset{(CH_2)_7\ CO_2Me}{\diagup}} \xrightarrow[HO(CH_2)_2OH/H_3O^+]{\text{Protection of } \!>\!C=O} \left[\begin{matrix}O \\ O\end{matrix}\right] C \underset{(CH_2)_7\ CO_2Me}{\overset{(CH_2)_7\ CO_2Me}{\diagup}} \longrightarrow$$

$$\xrightarrow[\text{Na/Xylene}]{\text{acyloin synthesis}} \left[\begin{matrix}O \\ O\end{matrix}\right] C \underset{(CH_2)_7-CHOH}{\overset{(CH_2)_7-CO}{\diagup}} \xrightarrow{H_2/Ni} \left[\begin{matrix}O \\ O\end{matrix}\right] C \underset{(CH_2)_7-CHOH}{\overset{(CH_2)_7-CHOH}{\diagup}} \longrightarrow$$

Acyloin

cis-and trans -forms

$$\xrightarrow[CH_3CO_2H]{HBr} O=C \underset{(CH_2)_7-CHBr}{\overset{(CH_2)_7-CHOAc}{\diagup}} \xrightarrow[\text{EtOH}]{\text{Zn}} O=C \underset{(CH_2)_7-CH}{\overset{(CH_2)_7-CH}{\diagup}}$$

cis-and trans-forms

Civetone
cis-and trans-forms

The *cis*- and *trans*- forms, referred to as α- and β- forms, were separated by fractional crystallisation of their dioxalans. The β- form was treated as follows to give the *cis*- civetone (or the α-form).

β-civetone

cis- Civetone

In the above transformation, the catalytic hydrogenation of a triple bond normally gives *cis*-addition, the configuration of the product is *cis* (α-civetone). This α- civetone was shown to be identical to natural civetone.

7.3 MUSCONE

It occurs in natural musk and was isolated from a scent gland of the Tibetan musk deer, and is used in preparations of purfumes (like civetone). Muscone is a thick colourless oil and is optically active. Its molecular formula is $C_{16}H_{30}O$. Muscone, like civetone was shown to contain a keto group. It was found to be monocyclic, since it was also shown to be saturated (an open chain compound will have the molecular formula $C_{16}H_{32}O$). If had contained more than one ring, the number of hydrogen atoms would have been less than 30.

The structure of muscone was established by Ruzicka as in the case of civetone. Ruzicka found that the odour of muscone was identical with that of synthetic cyclo-pentadecanone and its methyl derivatives. He, therefore, believed that muscone was a methyl derivative of a fifteen-membered cyclic ketone. This was proved by preparing cyclopentadecanone (by distillatio of the thoruim salt of tetradecane-1, 14-dicarboxylic acid), treating it with methyl magnasium iodide, and dehydrating the resulting tertiary alcohol to the alkene which was then catalytically reduced to methyl cyclopentadecane.

$$\begin{array}{c} CH_2 \!-\! CO \\ | \qquad | \\ (CH_2)_{12} \!-\! CH_2 \end{array} \xrightarrow[\text{(2) } H_2O]{\text{(1) } CH_3MgI} \begin{array}{c} \qquad CH_3 \\ \qquad | \\ CH_2 \!-\! C \!-\! OH \\ | \qquad | \\ (CH_2)_{12} \!-\! CH_2 \end{array} \xrightarrow{H_2O}$$

$$\begin{array}{c} \qquad CH_3 \\ \qquad | \\ CH_2 \!-\! C \\ | \qquad \| \\ (CH_2)_2 \!-\! CH \end{array} \xrightarrow{H_2 / Ni} \begin{array}{c} \qquad CH_3 \\ \qquad | \\ CH_2 \!-\! CH \\ | \qquad | \\ (CH_2)_{12} \!-\! CH_2 \end{array}$$

Methylcyclopentadecane is also obtained when muscone is reduced by Clemmensen method.

Now, the problem was to know the position of the methyl group with respect to the carbonyl group. On oxidation with chromic acid, muscone gives two acids of the molecular formula $C_{14}H_{28}$ $(CO_2H)_2$ and also a mixture of even-membered dicarboxylic acids ranging from succinic to decane-1, 10-dicarboxylic acid. This indicates that there is a straight chain of at least ten methylene groups in muscone. So muscone can be represented as I, II or III.

$$\begin{array}{c} CH_3 \!-\! CH(CH_2)_2 \!-\! CH_2 \\ | \qquad\qquad | \\ (CH_2)_{10} \!-\!\!-\!\! CO \\ \\ I \end{array} \qquad \begin{array}{c} CH_3 \!-\! CH\, CH_2 \!-\! CH_2 \\ | \qquad\qquad | \\ (CH_2)_{11} \!-\!\!-\!\! CO \\ \\ II \end{array} \qquad \begin{array}{c} CH_3 \!-\! CH \!-\!\!-\!\!-\! CH_2 \\ | \qquad\qquad | \\ (CH_2)_{12} \!-\!\! CO \\ \\ III \end{array}$$

Ruzick was unsuccessful to separating the mixture of two C_{14} dicarboxylic acids (see above) into two pure specimens. However, he obtained a partially pure specimens of one of them, which was shown to be identical to a synthetic specimen of 1-methyltridecane-1,13-dicarboxylic acid. So muscone has the structure III, with was confirmed by its synthesis (Ziegler, 1934).

$$\begin{array}{c} CH_3 \!-\! CH \!-\! CH_2CN \\ | \\ (CH_2)_{12}CN \end{array} \xrightarrow[\text{(2) } H_3O^+]{\text{(1) Li N(Ph) (Et)}} \begin{array}{c} CH_3 \!-\! CH \!-\!\!-\!\!-\! CH_2 \\ | \qquad\qquad | \\ (CH_2)_{12} \!-\! CO \end{array}$$

$$(\pm)\text{-muscone}$$

The formation of the two dicarboxylic acids $C_{14}H_{28}(COOH)_2$ from muscone by chromatic acid oxidation can be shown as follows:

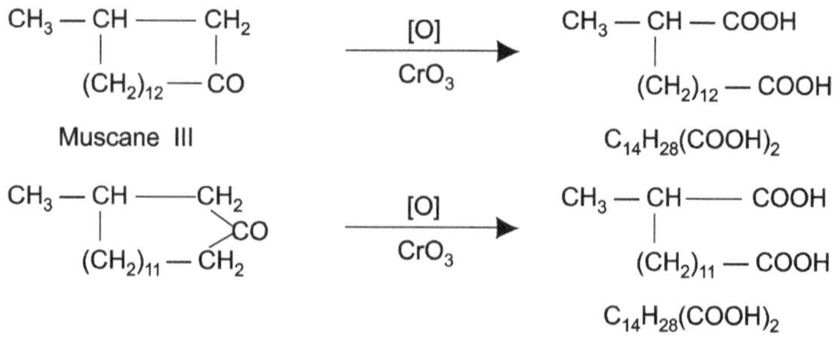
CH3 — CH — CH2
(CH2)12 — CO
Muscane III
[O]
CrO3
CH3 — CH — COOH
(CH2)12 — COOH
C14H28(COOH)2
CH3 — CH — CH2
CO
(CH2)11 — CH2
[O]
CrO3
CH3 — CH — COOH
(CH2)11 — COOH
C14H28(COOH)2

Muscone has also been obtained from dodecatrienyl nickel by reaction with allene and carbon monoxide followed by hydrogenation (R. Baker et. al., 1972).

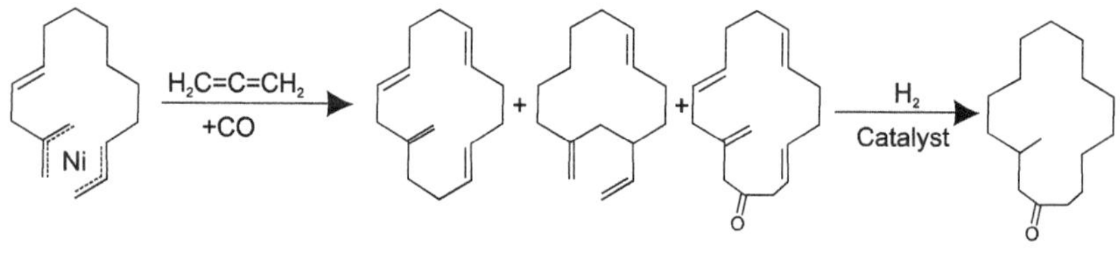
Ni
H2C=C=CH2
+CO
H2
Catalyst

Muscone

❑❑❑

8

Conformations of Cycloalkanes

Cycloalkanes, as we know, are cyclic compounds and contain carbon-carbon single bonds as in the case of alkanes. Examples include cyclopropane cyclobutane, cyclo pentane and cyclohexane. However, cycloalkanes having more than six carbons are also known.

The relative stabilities of cycloalkanes has already been discussed (see section 4.5.4).

8.1 CONFORMATIONS OF CYCLOPROPANE

In cyclopropane all the carbons are in one plane and the hydrogen atoms are situated below and above the plane of the ring and is a flat molecule. So there are no conformational isomers in cyclopropane [Fig. 8.1(a)].

In cyclopropane (a molecule with the shape of a regular triangle) the internal angle is 60° a value which is less by 49.5° from the tetrahedral angle of 109.5°. This results in **angle strain**. This is because the sp^3 orbitals of the carbon atoms are unable to overlap as effectively [Fig. 8.1(b)] as they do in alkanes. In fact, the carbon-carbon bonds of cyclopropane are described as 'bent'. The angle strain in cyclopropane is responsible for the ring strain. The hydrogen atoms of the ring are all eclipsed [Fig. 8.1(c)] and the molecule also has torsional strain. A Newman projection formula [Fig. 8.1(c)] as viewed along a carbon-carbon bond shows the eclipsed hydrogens.

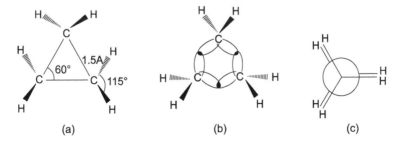

    (a)   (b)   (c)

Fig. 8.1 Angle strain in cyclopropane

© The Author(s) 2023
V. K. Ahluwalia and R. Aggarwal, *Alicyclic Chemistry*,
https://doi.org/10.1007/978-3-031-36068-8_8

8.2 CONFORMATIONS OF CYCLOBUTANE

Like cyclopropane, cyclobutane also has angle strain. In cyclobutane, the internal angles are 88°, a departure of more than 21° from the tetrahedral bond angle. Unlike cyclopropane, the cyclobutane is slightly folded. In case the cyclobutane ring is planar, the internal angles will be 90° instead of 88°, but the torsional strain will be considerably larger due to eight hydrogens being eclipsed. By bending slightly or folding, the cyclobutane ring relieves most of its torsional strain than it gains in the slight increase in its angle strain. The folded or bent conformations of cyclobutane are shown in Fig. 8.2

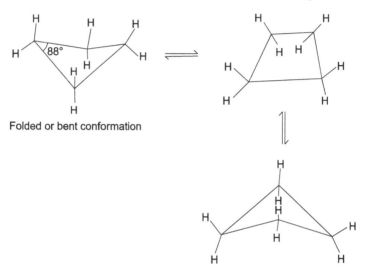

Folded or bent conformation

Fig. 8.2 Conformations of cyclobutane

The above non-planar structure of cyclobutane is supported by election diffraction measurements, x-ray data and the molecular dipole moments of cis- and trans forms of 1,3-dibromo or dicyano-cyclobutane(I, II and III resp.)

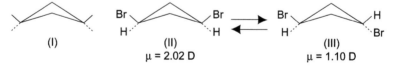

The *cis-* (IV) form of cyclobutane-1,3-dicarboxylic acid appears to have the pluckered ring while its *trans* isomer (V) appears to be essentially planar. On the basis of X-ray data, the carbon–carbon bond length in cyclobutane is slightly longer (156.7 pm) as compared to carbon–carbon bond length (154 pm) in cyclohexane.

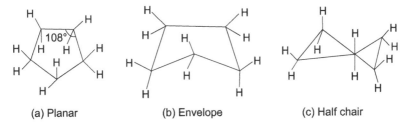

(IV) (V)

8.3 CONFORMATIONS OF CYCLOPENTANE

Like cyclobutane, cyclopentane has three conformations. A completely planar conformation of cyclopentane [Fig. 8.3(a)] has all C—H bonds eclipsed.

The internal angles of cyclopentane are 108°, a value close to the tetrahedral bond angle 109.5°. Thus, if cyclopentane molecule is planar, it would have little angle strain. However, planarity would introduce considerable torsional strain due to all 10 hydrogen atoms being eclipsed. Therefore, cyclopentane (like cyclobutane) assumes a slightly bent conformation (envelope conformation) in which two or more of the atoms of the ring are out of plane of the others [Fig.8.3(b)]. Due to this, some of the torsional strain is relieved. With little torsional strain and angle strain, the cyclopentane is virtually as stable as cyclohexane. A third conformation of cyclopentane is the half-chair conformation [Fig. 8.3(c)] which is nearly of equal energy as that the envelope; it has two atoms out of plane of the three others, one above and one below.

(a) Planar (b) Envelope (c) Half chair

Fig. 8.3 Conformations of cyclopentane

On the basis of combustion data, cyclopentane has a strain energy of about 27.3 kJ/mole and the torsional strain due to five eclipsed methylene units in a planar structure must be reduced by twisting of the ring. The thermodynamic properties indicate that cyclopentane has a pluckered ring. Also, the X-ray data of some natural products containing a cyclopentane ring indicate a bond angle of c. 105° (smaller than the tetrahedral angle).

In case two substituents are present in the cyclopropane ring, there is apossibility of cis-trans isomerism. Let us consider the example of 1,2-dimethylcyclopentane. It can be represented in two forms, cis 1,2-dimethylcyclopentane (the methyl groups are on the same side of the ring) and trans 1, 2-dimethylcyclopentane (the methyl groups are on the opposite side of the ring) (Fig. 8.4).

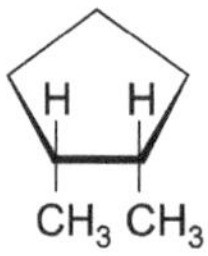

cis-1,2-Dimethylcyclopentane

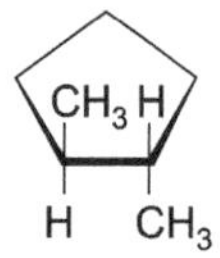

trans-1,2-Dimethylcyclopentane
b.p. 91.9°C

Fig. 8.4 cis and trans 1, 2-dimethylcyclopentane

The cis and trans 1, 2-dimethylcyclopentane are stereoisomers. These differ from each other in the arrangement of atoms in space. These forms cannot be interconverted without cleavage of carbon-carbon bond.

In a similar way, 1,3-dimethylcyclopentane also exhibit cis-trans isomerism (Fig. 8.5).

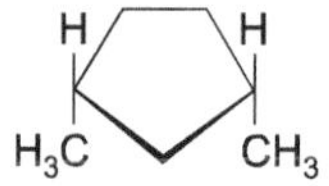

cis-1,3-Dimethylcyclopentane

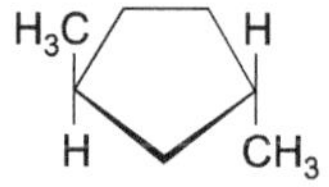

trans-1,3-Dimethylcyclopentane

Fig. 8.5 cis and trans forms of 1, 3-dimethylcyclopentane

The heats of combustion indicate that cis-1,3-dimethylcyclopentane is more stable than the trans- form by about 2.1 kJ/mole; this led to the suggestion that 1,3-dimethylcyclopentane has the conformation (A) (called the envelope conformation with four carbon atoms in one plane and one outside the plane).

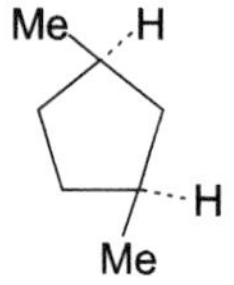

cis-1,3-Dimethyl
cyclopentane

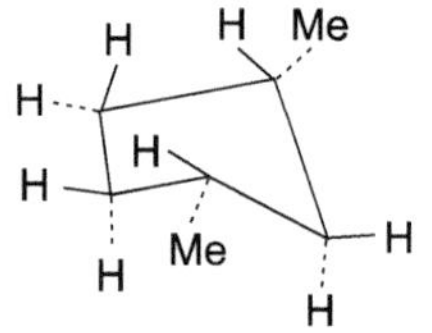

(A)

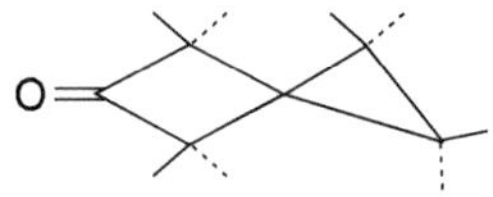

halt-chair
conformation

The effect of twisting a particular $-CH_2-$ in cyclopentane is transmitted around the ring in a sequence that amounts to a process of pseudo rotation. However, in cyclopentane, the triagonal centre in the molecular structure removes a number of eclipsing bonds, and a half chair conformation (with three carbon atoms in one plane and the other two outside.) rather than the envelope conformation (A) becomes the preferred form.

8.4 CONFORMATIONS OF CYCLOHEXANE

Cyclohexane is known to exist in two main conformations known as the chair and boat conformations. The chair conformation of cyclohexane is most stable. It is a non-planar structure, the carbon-carbon bond angles are all 109.5°. The chair conformation is free of angle strain and also of torsional strain. On viewing along any carbon-carbon bond, the atoms are perfectly staggered and the hydrogen atoms at opposite corners of the cyclohexane ring are maximally separated (see Fig. 8.6)

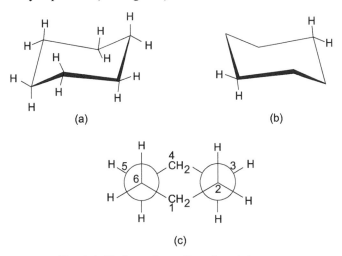

(a)   (b)

(c)

Fig. 8.6 Chair conformation of cyclohexane

(a) Line drawing, (b) Illustration of large separation between H atoms at opposite corners and (c) Newmann projection of chair conformation.

The boat conformation of cyclohexane is formed by flipping one end of the chair form (upward or downward). The flipping requires only rotation about

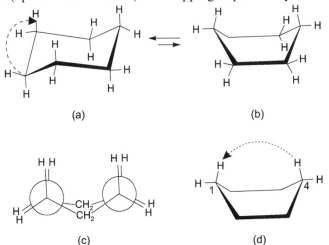

(a)   (b)

(c)   (d)

Fig. 8.7 Boat conformations of cyclohexane (a) Chair conformation and (b) its conversion into boat conformation (c) Newmann projection of boat conformation (d) Flagpole interaction of the C1 an C4 hydrogen atoms of boat conformation

carbon-carbon single bond. The boat conformation like the chair conformation is also free of angle strain. However, the boat conformation is not free of torsional strain (see Fig. 8.7))

As seen the H substituents in the boat conformation are eclipsed [see Fig. 8.7(c)], Newmann Projection). Also the two hydrogens on Cl and C4 are close enough to cause van der Waals repulsion [Fig. 8.7 (d)]; this effect is called the 'flagpole' interaction of the boat conformation. Due to torsional strain and flagpole interactions, the boat conformation has higher energy than the chair conformation.

The conversion of boat form into the chair form involves the pulling down of the uppermost carbon down so that it becomes the bottom-most carbon atom. However, in case the uppermost carbon is pulled just a little, another conformation, known as twist boat or skew boat conformation results (Fig. 8.8). During this conversion, some of the torsional strain of boat conformation is relieved.

Fig. 8.8 Twist boat conformation of cyclohexane

The various conformations of cyclohexane are chair, half-chair and twist boat and boat conformation. The interconversion of these formation are shown in Fig. 8.9.

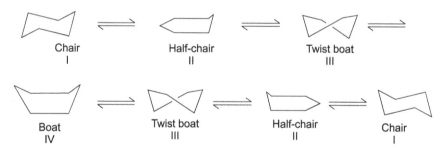

Fig. 8.9 Interconversion of conformers of cyclohexane.

The relative energies of various conformations of cyclohexane are depicted in Fig. 8.10.

As seen (Fig. 8.10), the energy barriers between the chair, boat and twist boat conformations of cyclohexane are low. This makes the separation of the conformations rather impossible at room temperature.

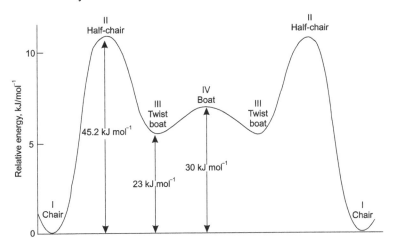

Fig. 8.10 Relative energies of the various conformations of cyclohexane

At room temperature about 1 million interconversions occur each second and due to greater stability of the chair, more than 99 per cent of the molecules are believed to be in the chair conformation at any time.

The NMR spectra of cyclohexane shows a sharp peak at δ1.5–2.0 (the region of methylene absorption), due to the fact that the two types of protons are indistinguishable due to rapid reversible interconversion of one form of cyclohexane into another. However, at the temperature, when its equilibrium is frozen, two signals are observed — one upheld (due to axial protons) and the other downfield (due to equatorial protons).

For the development and applications of the principles of conformation in chemistry, Derek H. R. Barton and Odd Hassel shared the Nobel Prize in 1969. Their work was useful for the understanding of not only the conformations of cyclohexane rings, but also the structures of other compounds (like steroids) containing cyclohexane rings.

8.4.1 Axial and Equatorial Bonds in Cyclohexane

The chair form of cyclohexane has two types of carbon-hydrogen bonds, viz., axial (a) and equatorial (e). The C—H bonds which are parallel to the axis are axial Bonds (Fig. 8.11) and the remaining six C—H bonds which extend outward at an angle of 109.5° to the axis are equatorial bond (Fig. 8.12). The axial bonds are vertical to the plane of the ring and the equatorial bonds are roughly in the plane of the rings.

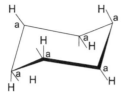 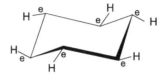

Fig. 8.11 Axial C—H bonds in   **Fig. 8.12** Equatorial C—H bonds in
            cyclohexane   cyclohexane

When ring flipping takes place from one chair form to another, all axial bonds become equatorial bonds and equatorial bonds become axial bonds.

In case of substituted cyclohexanes (mono or di) there will be two cyclohexanes, axial or equatorial. It is interesting to know why one form is more stable than the other. This forms the subject-matter of the following sections.

8.4.2 Conformations of Monosubstituted Cyclohexane

Monosubstituted cyclohexane, for example, methylcyclohexane has two chair conformations. In one form methyl group is axial and in the other the methyl group is equatorial (Fig. 8.13). Both these forms are interconvertible by ring flipping.

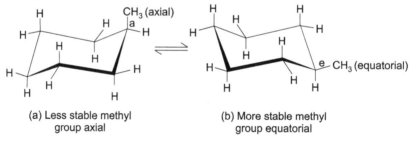

(a) Less stable methyl   (b) More stable methyl
    group axial   group equatorial

Fig. 8.13 Conformations of methylcyclohexane

On the basis of studies, it has been shown that methyl cyclohexane with the methyl group equatorial is more stable than the conformation with methyl group axial by about 7.6 kJ mol^{-1}. In fact the methylcyclohexane with equatorial methyl is present to the extent of 95 per cent in the equilibrium mixture. The greater stability of methylcyclohexane with equatorial methyl is because in axial methyl group (in position 1), there is non-bonded interaction between CH_3 group and H atoms at positions 1 and 3, commonly known as **1,3-diaxial interaction**. In this case the distance between H of CH_3 and H at C-3 and C-5 is less than the sum of van der Waals radii of two hydrogens. This repulsion destabilises axial conformation of methylcyclohexane whereas no such interaction is possible with equatorial conformation of methylcyclohexane making it more stable conformation (Fig. 8.14).

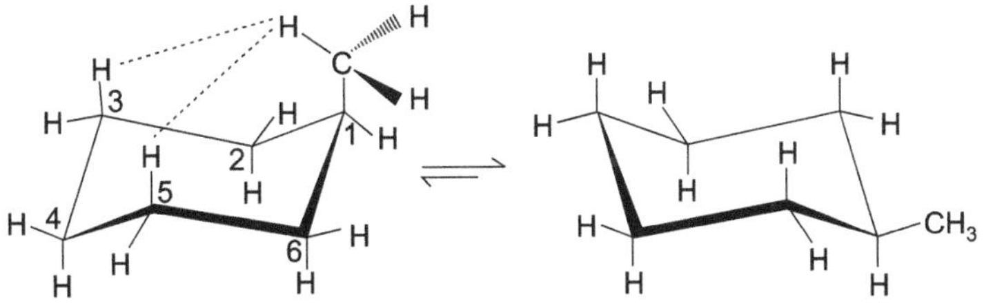

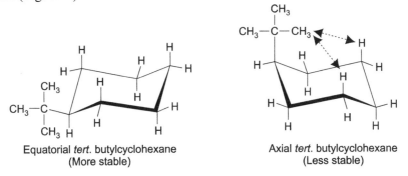

1,3-diaxial interactions
(Less stable)

No 1,3-diaxial interactions
(More stable)

Fig. 8.14 Equatorial methylcyclohexane is more stable than axial methylcyclohexane

In case monosubstituted cyclohexane has a large alkyl substitutent (e.g., tertiary butyl group), the strain caused by 1,3-diaxial interaction is much more pronounced. The conformation of *tert*-butylcyclohexane with *tert*-butyl group in equatorial is found to be more than 21 kJ mol^{-1} more stable than the axial form (Fig. 8.15).

Equatorial *tert.* butylcyclohexane
(More stable)

Axial *tert.* butylcyclohexane
(Less stable)

Fig. 8.15 Equatorial *tert.* butylcyclohexane is more stable than axial *tert.* butylcyclohexane

At room temperature 99.99 per cent of is equatorial *tert*-butylcyclohexane is present in the equilibrium mixture.

8.4.3 Conformations of Disubstituted Cyclohexanes

8.4.3.1 Conformations of 1,2-Dimethylcyclohexane

As in the case of 1,2-dimethylcyclopentane in case of 1,2-dimethylcyclohexane also there is a possibility of cis-trans isomerism.

Let us first consider the case of 1,2-dimethylcyclohexane, in which one CH_3 group occupies axial position at C-1 and the second methyl group occupies an equatorial position at C-2. As both the methyl groups are on the same side, this is the cis arrangement (Fig. 8.16).

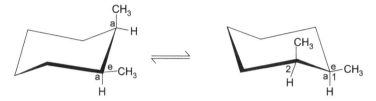

Fig. 8.16 *Cis* 1,2-Dimethylcyclohexane

In the above case, the two CH_3 groups are closer to each other, there is a crowding between the hydrogens of the two methyl groups. Ring flipping of the cis form gives another equivalent cis form. However, this does not lead to any change as far as interactions between the hydrogens of the two methyl groups are concerned.

In case both the methyl groups occupy axial position, the arrangement is trans arrangement of the groups. This trans conformation on ring flipping gives another trans configuration in which both the methyl groups occupy equatorial positions (Fig. 8.17).

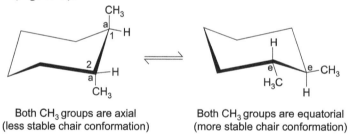

Both CH_3 groups are axial   Both CH_3 groups are equatorial
(less stable chair conformation)   (more stable chair conformation)

Fig. 8.17 *Trans* 1,2-Dimethylcyclohexane

In the above configuration, the axial —CH_3 group at C-1 faces van der Waals repulsion by axial hydrogens at C-3 and C-5 carbon atoms. Similar repulsion for the axial C-2 methyl group with hydrogens at C-4 and C-6 carbon atoms, make this diaxial trans conformation less stable compared to the diequatorial trans conformation. The equatorial positions are free from such interactions as the substituents project outward. As a general rule, any substituent is more stable in the equatorial position than in the axial position. Thus, trans diequatorial conformation of 1,2-dimethylcyclohexane is more stable than the trans diaxial cyclohexane.

8.4.3.2 Conformations of 1,3-Dimethylcyclohexane

In 1,3-dimethylcyclohexane, when both the methyl groups are axial, the arrangement is cis. This arrangement on ring flipping gives another cis form in which both the methyl groups occupy equatorial positions (Fig. 8.18).

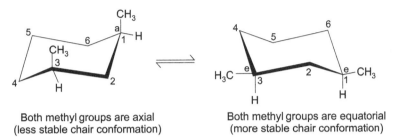

Both methyl groups are axial   Both methyl groups are equatorial
(less stable chair conformation)   (more stable chair conformation)

Fig. 8.18   *Cis* 1,3-Dimethylcyclohexane

However, in case C-1 methyl is axial and C-3 methyl is equatorial (*i.e.*, both the CH_3 groups are trans to each other), ring flipping gives another equivalent trans form (Fig. 8.19).

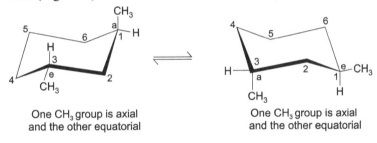

One CH_3 group is axial   One CH_3 group is axial
and the other equatorial   and the other equatorial

Fig. 8.19   *Trans* 1,3-Dimethylcyclohexane

As seen, the *trans* form has one methyl group in equatorial position and the other methyl group in axial position compared to the cis form, which is its more stable conformation both the methyl groups are in equatorial positions. So in this case, the cis form with diequatorial substituents is more stable than the trans form.

8.4.3.3 Conformations of 1,4-Dimethylcyciohexane

In 1,4-dimethylcyclohexane, when both the methyl groups are axial, it leads to trans arrangement of methyl groups. The resulting diaxial arrangement of methyl groups on ring flipping is changed into another chair form having diequatorial arrangement of methyl groups (Fig. 8.20).

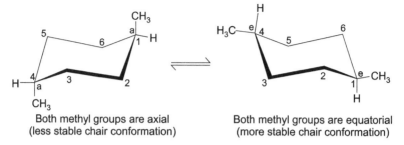

Both methyl groups are axial   Both methyl groups are equatorial
(less stable chair conformation)   (more stable chair conformation)

Fig. 8.20 *Trans* 1,4-Dimethylcyclohexane

If one methyl at C-1 is axial and the other methyl at C-4 is equatorial, it leads to cis arrangement of methyl groups. The ring flipping gives another equivalent chair conformation (Fig. 8.21).

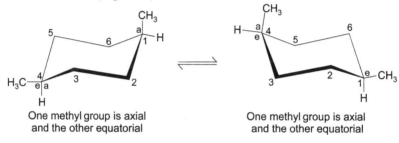

One methyl group is axial   One methyl group is axial
and the other equatorial   and the other equatorial

Fig. 8.21 *Cis* 1,4-Dimethylcyclohexane

The cis-1,4-dimethyl cyclohexane (Fig. 8.21) has two equivalent chair conformations having one axial and equatorial methyl substituent.

We have seen that in cycloalkanes, like cyclopropane, cyclobutane, cyclopentane and cyclohexane, if there are two substituent, these can be *cis-* or *trans*.

Comparison of the cis and trans conformations shows that in the trans conformation both the methyl groups are equatorial and so is more stable than the cis form which has one substituent each in the axial and equatorial positions.

The different conformations of 1,2-, 1,3- and 1,4-dimethylcyclohexanes are summarised in the Table given below. The more stable conformation, where one exists, is shown in bold type.

Table Conformations of dimethylcyclohexane

Compound	cis isomer	Trans isomer
1,2-dimethylcyclohexane	a, e or e, a	**e, e** or a, a
1,3-dimethylcyclohexane	**e, e** or a, a	a, e or e, a
1,4-dimethylcyclohexane	a, e or e, a	**e, e** or a, a

As seen the conformation in which both the methyl groups are in equatorial positions are more stable.

In case, the disubstituted cyclohexanes have different substituent, the isomer with the larger substituent occupying the equatorial position is more stable.

Cyclohexane-1,3-diol, however, conform to the diaxial rather than the diequatorial orientation because intramolecular hydrogen bonding, possible only in diaxial form, stabilises the diaxial form.

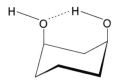

Similarly, in *trans*-1,2-dibromocyclohexane, the diaxial conformation makes more contribution as the C – Br dipoles are opposed to each other.

8.4.4 Cyclohexene

The four groups attached to a double bond in cyclohexene must lie in the same plane in order to provide the most effective overlap for π bond. In cyclohexene, this constraint gives a structure which is a half chair to provide effective staggering of the four $-CH_2-$ groups and thus has little strain. X-ray data also show that cyclohexene ring preferentially adopts a half chair conformation (22). In the NMR spectrum of deuterocyclohexene at $-170°C$, distinct signals are observed for the proton (marked with asterisk) in the two conformations (23 a and b). However, at higher temperatures, an average signal is observed beyond 8.4.

(22)

(a)
(23)
Deuterocyclohexene

The heat of hydrogenation of cyclohexene is 28.4 kcal/mol, which is comparable with that of cis-2-butene. Cyclohexene, thus, behaves as a typical *cis*-alkene, and via *trans* addition first gives the diaxial cyclohexyl product which then rapidly converts to the corresponding diequatorial confomer.

8.4.5 Cyclohexanone

The cyclohexanone ring is buckled slightly from the chair conformation in order to accomodate the trigonal carbon atom whose optimal angle is 120°. The equatorial hydrogen atom on the α-C atom becomes nearly eclipsed by the carbonyl oxygen; thus, the equatorial substitutions are somewhat destabilised by steric repulsions. In some cases, this leads to a greater stability of the axially substituted conformer, *e.g.*, 2-bromo-cyclohexnone.

Infra-red spectra studies have shown that the bromine in 2-bromocyclohexanone is predominantly axial. The C – Br and C=O bonds are both strongly polar and when the bromine is equatorial, the dipolar repulsion is maximum; it is minimum when the bromine is axial. Since the axial form predominates, this equatorial dipolar repulsion must be larger than the 1,3-diaxial interactions. When, however, other substituents are present, the 1,3-interaction may become so large as to overweigh the dipolar effect and the bromine would now be equatorial.

8.4.6 Cyclohexane Structures having Boat Conformations

We have now known that in a majority of cyclohexane derivatives, the preferred conformation is the chair form. However, in certain situations this is reversed, either by the presence of a very large axial substituent (as in *trans*-1,3-*ditert*.butylcyclohexane) or by the presence of one or more trigonal centres.

In the first case, the *trans*-1,3-*ditert*.butylcyclohexane may be equilibrated by heating with palladium catalyst, and its two forms are represented by (24a) and (24b).

(a)   (b)

(24) The chair (a) and boat (b) conformations of *trans*-1,3-*ditert*.butylcyclohexane

In order to avoid axially oriented *tert*. butyl group, the 1,3 trans-isomer must necessarily take up the boat conformation (24b). From the change in ratio (24a)/(24b) with temperature, the difference in free energy and enthalpy between the boat and chair conformations may be derived. From the heats of combustion and other data, it is found that for chair → boat conversion, ΔH is +26 kJ/mole and ΔS is + 12.2 J/deg mole. The chair → boat inversion is characterised by a large positive entropy change since the chair is rigid and the twisted boat is a flexible structure.

In the second case, the best example is that of cyclohexane-1,4-dione. X-ray analysis shows that in solid state, the preferred conformation is (25), with an angle of 154° between the two >C=O bonds, and a net molecular dipole moment of about 1 Debye unit.

$$O = \hspace{4cm} = O$$

(25) The Preferred conformation of cyclohexane-1,4-dione

When intramolecular hydrogen bonding is possible between groups in the 1 and 4 positions, the cyclohexane ring may assume a boat conformation. A twist boat form may also exist as a varint.

Boat conformation   Twist boat form

Even when compounds exist predominantly in the chair form, the boat form may be of importance in some reactions, *e.g.*, the lactonisation of *cis*-

4- hydroxycyclohexane carboxylic acid which must proceed *via* the boat conformation; the *trans* form does not lactonise.

We encounter with boat type of conformations in the derivatives of both *cis*- and *trans*-decalins (to be discussed subsequently). Thus, in the steroid ketone (26), though there is dipole repulsion between the C=O and C–Br bonds and a trigonal group is present at C_3, the energy of conversion of chair → boat is reduced.

(26)

Similar is the case with the steroid ketone (27).

(27)

8.5 CONFORMATIONS OF CYCLOHEPTANE

Cycloheptane exists in non-planar conformation (free of any angle strain). However, there is no rigid and preferred conformation as in case of cyclohexane (chair conformation). In the chair-type conformation (28a), there is an eclipsed ethane unit at positions 4,5 and the hydrogens marked with asterisks, at positions 3 and 6 are 128 pm apart. The ring, however, is flexible and a twist chair form (28b) is possible which reduces the torsion at 4,5 and increases the distance at the 3,6-hydrogens to 186 pm. At low temperature (–170°C), the two fluorine substituents in 1,1-difluorocycloheptane exhibit no chemical shift difference, indicating very rapid conformational cange.

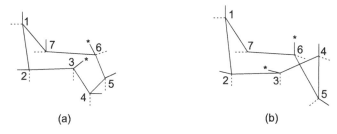

(a)   (b)

(28) Cycloheptane (a) chair-type conformation; (b) twist chair form

In cycloheptene, the olefinic bond makes the system rigid and the available evidence points to a preferred chair conformation (29). The conformational preference (29) is supported by its low temperature NMR spectrum which shows splitting at the signals for the CH_2^* protons at about $\tau 8.75$.

(29)

8.6 CONFORMATIONS OF CYCLOOCTANE

Cyclooctane like cycloheptane also exists in a non-planar conformation (free of angle strain). In this case also, there is no clearly preferred conformation. The X-ray analysis of cyclooctane-*trans*-1,2-dicarboxylic acid and other cyclooctane derivatives establishes a chair-boat conformation (30a); this is also established by the low temperature NMR spectrum of deuterocyclooctanes. The ring, however, is flexible and both *tert.* butylcyclooctane (30b) and 5-*tert*.butyl-cyclooctanone (30c) show two unequal NMR bonds due to *tert.* butyl group. This implies that there is more than one conformtion.

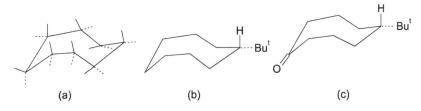

(a)   (b)   (c)

30 Chair-boat conformation of cycloheptane
(a) Cyclooctane derivatives (b) 5-*tert*.butylcyclooctane and
(c) 5-*tert*.butylcyclooctanone

8.6.1 Conformation of Cyclodecane

Cyclodecane like cycloheptane and cyclooctane exist in non-planar configuration. Its smaller unstability is caused by torsional stress and van der Walls repulsions between hydrogen atoms across rings, called **transannular strain**. The non-planar conformation is free of angle strain. X-ray crystallographic studies of cyclodecane reveal that the most stable conformation has carbon-carbon-carbon bond angle of 117°, indicating some strain. The wide bond angle, of course, permits the molecule to expand thereby minimizing unfavourable repulsions between hydrogen atoms across the ring (Fig. 8.31).

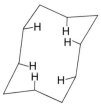

Fig. 8.31 Cyclodecane (H—H repulsions across the ring)

8.6.2 Conformations of Fused Six-membered Rings

Common example of fused six-membered rings contains a bicyclic system. An example of fused bicyclic six-membered rings is decalin, which is known to exist in two forms, viz., *cis-* and *trans* decalins (Fig. 8.32).

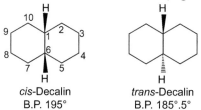

cis-Decalin
B.P. 195°

trans-Decalin
B.P. 185°.5°

Fig. 8.32 *Cis* and *trans* decalins

Decalin is the common name. It is described as bicyclo (4.4.0) decane A knowledge of the conformations of 1,2-dimethylcyclohexane is useful for the conformational analysis of fused saturated ring. As an example, trans-decalin can be visualised to be derived from most stable conformation of *trans*-1,2- dimethylcyclohexane by extending the methyl groups with two additional carbons into another ring. Alternatively, the two methyl groups can be visualised to be the two ends of a four carbon bridge, that is, —CH$_2$—CH$_2$—CH$_2$—CH$_2$— (see Fig. 8.33).

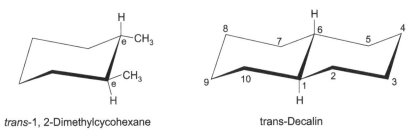

trans-1, 2-Dimethylcycohexane   trans-Decalin

Fig. 8.33 *Trans*-Decalin visualised to be derived from
trans-1,2-dimethylcyclohexane.

The *trans* ring fusion is understood by the relative positions of H atoms at the bridge head positions (the carbon atoms common to both the rings). However, unlike *trans*-1,2-dimethylcyclohexane, trans-decalin cannot flip into another stable chair-chair form.

In a similar way cis-decalin can be visualised to be formed from *cis*-1,2-dimethylcyclohexane (one CH_3 group axial and the other equatorial). In this isomer, chair-chair ring flipping is easy (Fig. 8.34).

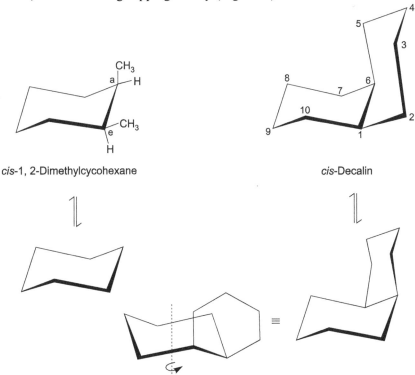

cis-1, 2-Dimethylcycohexane   *cis*-Decalin

Fig. 8.34. *Cis*- Decalin visualised to be derived from *cis*-1,2-dimethyl cyclohexane and its ring flipping to give another chair-chair conformation

The *trans-* and *cis*-fusion of chair cyclohexane units in decalins do not involve any change in dihedral angle. However, *cis*-decalin has relatively higher enthalpy due to more severe non-bonded interactions in its *cis*-form than in the *trans*-form.

In case there is a methyl group in position 6, it introduces increased non-bonded interaction; this is more serious in *trans*-isomers. Also, a trigonal centre such as carbonyl group, on the other hand, reduces axial interactions, which are more serious in the *cis*-decalin series. Table below gives the enthalpy differences for *cis* and *trans* forms of various fused ring structures.

Table Enthalpy differences for cis and trans forms of various
fused ring structures

Compounds			Enthalpy differences kJ/ mole
Decalin	R = R' = H	*trans → cis*	11.5
10-Methyidecalin	R = CH₃; R' = H	*trans → cis*	5.9
10-Methyl-1-decalone,	R = H; R' = O	*trans → cis*	0.8
Hydrindane	(structure given below)	*cis → trans*	4.2
Perhydroazulene	(structure given below)	*cis → trans*	small
Bicydo[3,3,0]octane	(structure given below)	*cis → trans*	29.6

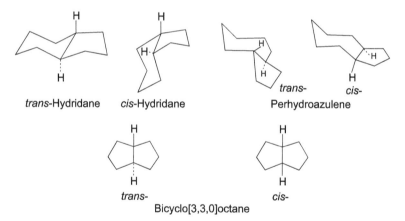

trans-Hydridane   cis-Hydridane   Perhydroazulene

trans-   cis-
Bicyclo[3,3,0]octane

Some other examples of bicyclic systems include the following:

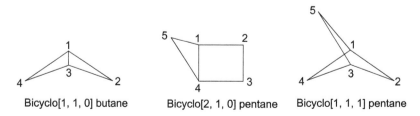

Bicyclo[1, 1, 0] butane   Bicyclo[2, 1, 0] pentane   Bicyclo[1, 1, 1] pentane

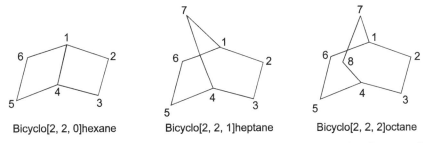

Bicyclo[2, 2, 0]hexane   Bicyclo[2, 2, 1]heptane   Bicyclo[2, 2, 2]octane

The conformation of the fused ring system is of particular interest in determination of the stereochemistry of steroids. A typical example is that of cholesterol.

It has been found that in cholesterol.

Rings A and B are *cis* fused.

Rings B and C are *trans* fused and

Rings C and D are *trans* fused.

So far we have seen conformations of bicyclic rings. Let us now consider a tricyclic system containing a three-dimensional cyclohexane ring [Adamantane, tricyclo $(3.3.1.1^{3,7})$ decane].

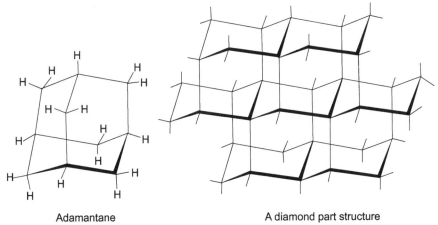

Adamantane   A diamond part structure

Fig. 8.35 Adamantane and diamond.

The three dimensional structure of diamond is derived from the structure of adamantane (Fig. 8.35).

The conformation of six-membered ring is useful for the structure elucidation of complex natural products like cholesterol.

A twist boat of adamantane is twistane. In twistane, the six membered ring (chair conformation) is forced into twist-boat conformation by the bridging carbon atoms.

Twistane

An example each of a tetracyclic and pentacyclic compounds are given below.

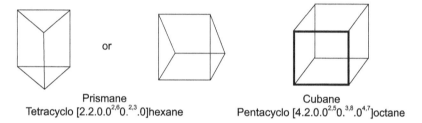

or

Prismane
Tetracyclo [2.2.0.$0^{2,6}$.$0^{2,3}$.0]hexane

Cubane
Pentacyclo [4.2.0.$0^{2,5}$.$0^{3,8}$.$0^{4,7}$]octane

Cyclooctatetraene is found to exist as bicyclo [4.2.0] -2,4,7-octatriene. This has been shown on the basis of adduct formation with maleic anhydride.

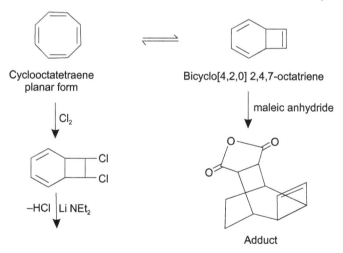

Cyclooctatetraene
planar form

Bicyclo[4,2,0] 2,4,7-octatriene

maleic anhydride

Adduct

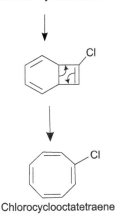

Chlorocyclooctatetraene

Adduct of cyclooctatetraene with maleic anhydride

8.7 CYCLOALKENES

Nomenclature

The nomenclature of cycloalkenes have been discussed earliver [See page 4]

Synthesis

Some of the methods for the synthesis of cycloalkenes, viz, cyclopropene, cyclobutene, cyclopentene and cyclohexene are given below.

Cyclopropene is obtained by the Hofmann elimination as given below.

$$\triangle\!\!-N^+Me_3\bar{O}H \xrightarrow[\text{Pt catalyst}]{300°C} \triangle$$

Cyclopropene

Cyclopropene derivatives can also be obtained by making use of the principle of carbene addition (see page 29)

Cyclobutene is obtained by the Hofmann or cope elimination reaction.

Cyclobutene

Cyclopentene is obtained by heating vinyl cyclopropane. This arrangement could be facelitated by allylic stabilization in a radical intermediate (route A) or 1,3-concertul migration as in route B (See page 31)

Vinyl Cyclopropene   Cyclopropentene

Cyclohexene can be prepared by (4 + 2) cycloaddition (Diel's alder reaction) of a conjugated diene (butadiene) (4π election) and an alkene (ethene, 2π electron).

Butadiene   Cyclohexene

8.7.1 Stereochemistry of Cycloalkenes

Cycloalkenes, like alkenes also contain a double bond. Cycloalkenes up to five carbon atoms exist only in the cis form. In cycloalkenes, introduction of a trans double bond will introduce much greater strain than the bonds of the ring system can accommodate.

Cyclopropene   Cyclobutene   Cyclopentene

cis cycloalkenes

The cyclohexene (A) might resemble the structure (B). It is believed that the structure (B) can be formed as a very reactive short-lived intermediate in some chemical reactions.

(A)   (B)

Cyclohexene

The structure (B) is highly strained to exist at room temperature.

In case of cycloheptene, only the trans form has been observed spectroscopically. It is a substance with a very short life time and it is not possible to isolate it

Cyclooctene can exist in both cis and trans forms. In case of cyclooctene the ring is large enough to accommodate the geometry necessary for a trans double bond. *Trans*-cyclooctene is stable at room temperature (unlike other cycloalkenes). However, trans-cyclooctene is chiral and exists as a pair of enantiomers.

cis-cyclooctene   trans-cyclooctene

cis and trans cyclooctene

Cycloheptatriene molecule is non-planar with rapid inversion between the following two equivalent structures A and B.

(A)   (B)

(Two equivalent structures of cycloheptatriene)

The inversion at low temperature is seen by NMR proton signals for Ha and H_b. (which are resolved) at low temperature.

Cycloheptatriene also exhibits thermal isomerism due to 1,5-hydrogen transfer across the molecule. This is supported on the basis of change in NMR signal for allylic and vinylic protons.

Thermal isomerism in cycloheptatriene

Cyclooctatetraene has been shown to exist in tub form with alternating single and double bonds. The bonds of cyclooctatetraene are shown to be alternatively long (1.48 A) and short (1.34 A) respectively (on the basis of X-ray studies).

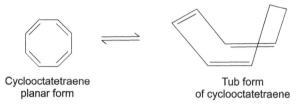

Cyclooctatetraene   Tub form
planar form   of cyclooctatetraene

Cyclooctatetraene

❑❑❑

Cycloalkanes Containing an Heteroatom (Heterocyclic Compounds)

9.1 INTRODUCTION

The cycloalkanes, as we know are made up of only carbon and hydrogen. These are known as homocyclic compounds. The replacement of one CH_2 group in cycloalkanes with an heteroatom like oxygen, nitrogen or sulphur gives cycloalkane derivatives containing an heteroatom. Such compounds are known as heterocyclic compounds.

The heterocyclic compounds contain a three, four, five or six membered ring system.

The three membered heterocyclic compounds are considered to be derived from cyclopropane by replacing a CH_2 group by a hetero atom (O, N or S). These are oxirane (epoxide), aziridine and thrirane.

The unsaturated analogues are

© The Author(s) 2023
V. K. Ahluwalia and R. Aggarwal, *Alicyclic Chemistry*,
https://doi.org/10.1007/978-3-031-36068-8_9

The four membered heterocyclic compounds are considered to be derived from cyclobutane by replacing a CH_2 group by a hetero atom (O, N or S). These are oxetane azetidine and thietane.

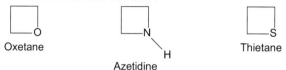

Oxetane   Azetidine   Thietane

The simplest five membered heterocyclic compounds containing one heteroatom are Furan, pyrrole and thiophene.

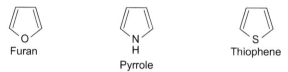

Furan   Pyrrole   Thiophene

Some of the typical six membered include α-pyran, γ-pyran and pyridine.

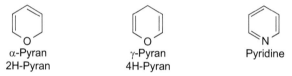

α-Pyran   γ-Pyran   Pyridine
2H-Pyran   4H-Pyran

9.1.1 Importance of Heterocyclic Compounds

The importance of heterocyclic compounds is apparent from the fact that different types of such compounds are of great importance for life processes. Such compounds include carbohydrates (which are classified as oxygen heterocycles), nucleic acids and some aminoacids (e.g., proline, hydroxyproline and tryptophan), peptides and proteins (containing nitrogen ring systems.

$C_6H_5CH_3$
Toluene

$\xrightarrow[\text{Crown ether}]{KMnO_4/\overline{O}H}$

C_6H_5COOH
80%
Benzoic acid

$\xrightarrow{KMnO_4}$

18 crown-6   Complex of crownether with $KMnO_4$

Some heterocyclic compounds, e.g., crown ethers (which are large ring compounds containing several oxygen atoms) are examples of heterocyclic compounds. A typical example is 18-crown-6, which is useful for the oxidation of toluene with $KMnO_4$ in alkaline medium to give benzoic acid in 80% yield.

Another heterocyclic compound encountered in nature is chlorophyll, which is present in green leaves and acts as a catalyst for photosynthesis. Chlorophyll has the porphyrin ring system.

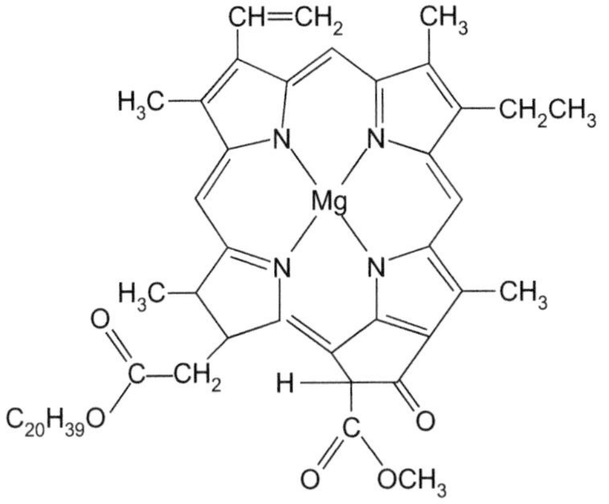

Chlorophyll a green plant pigment
catalyst for photosynthesis

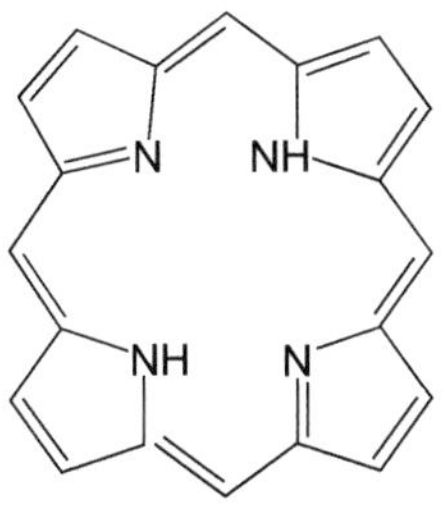

Porphyrin ring system

A number of heterocyclic compounds are used as medicine. Some examples are given below.

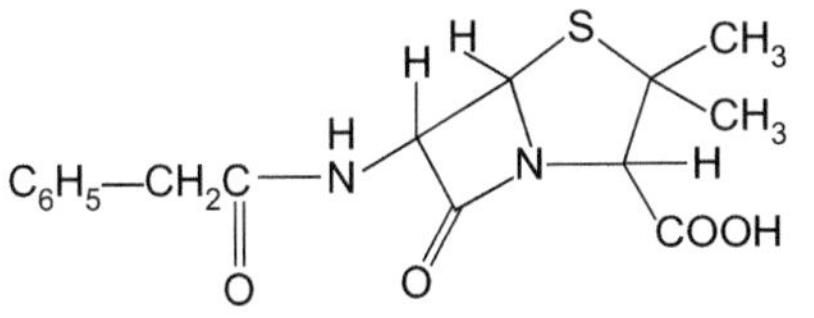

Penicellin y (Antibiotic)

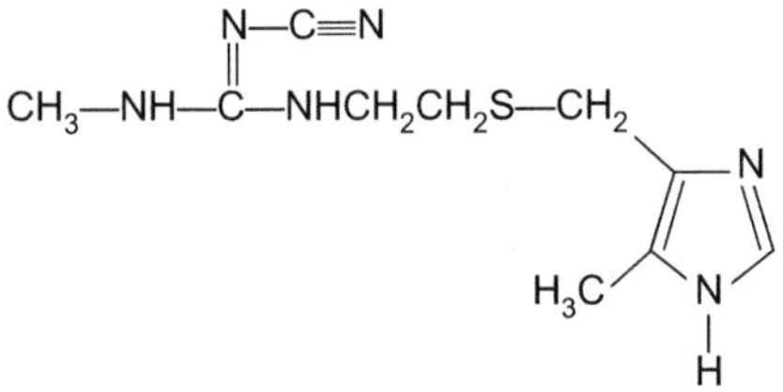

Cimetidine (a drug)

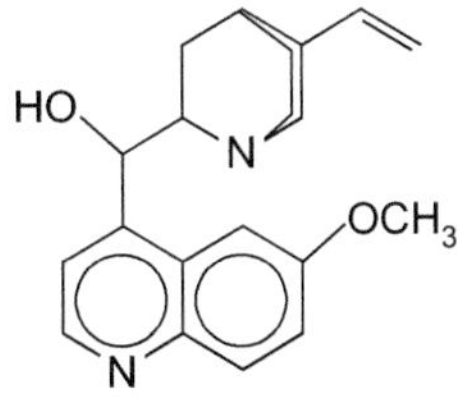

Quinine (Antimalarial)

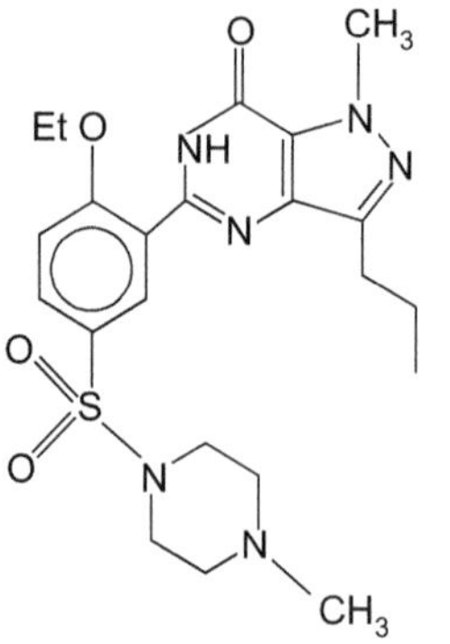

Viagra
(For treatment of male impotence)

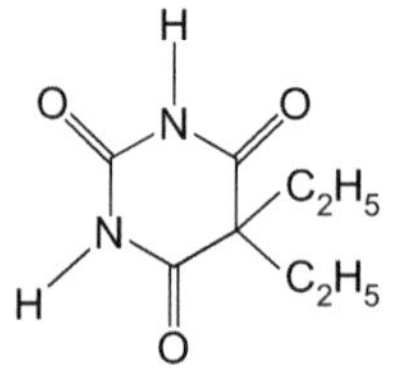

Phenobarbatal
(Sedative and hypnotic)

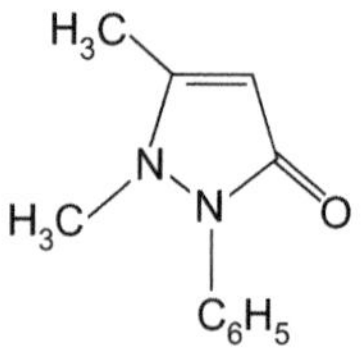

Antipyrene or phenazone
(For lowering body temperature

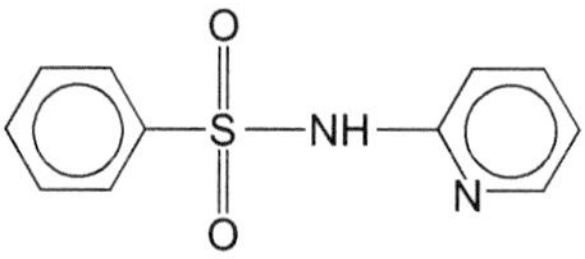

Sulfapyridine (antibiotic)

Some amino acids, essenial for life processes also contain a heterocyclic moiety. Some examples are given below.

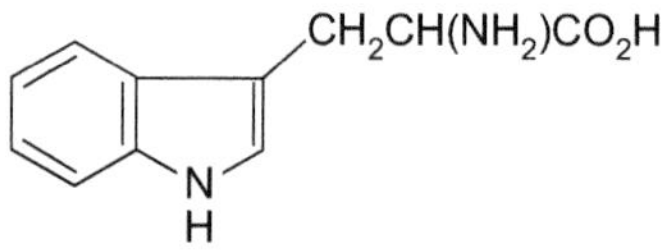

CH$_2$CH(NH$_2$)CO$_2$H

COOH

Tryptophan 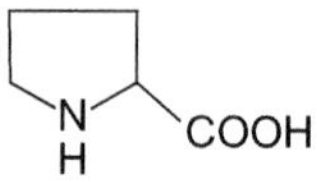   **Proline**

CH$_2$CH(NH$_2$)CO$_2$H

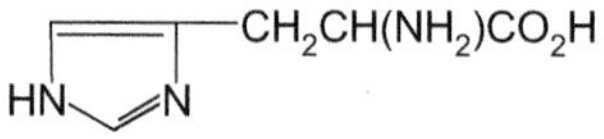

Histidine

A number of alkaloids and vitamins are built up of heteroxyelic moiety. Some examples are given below.

CH$_2$CH$_2$OH

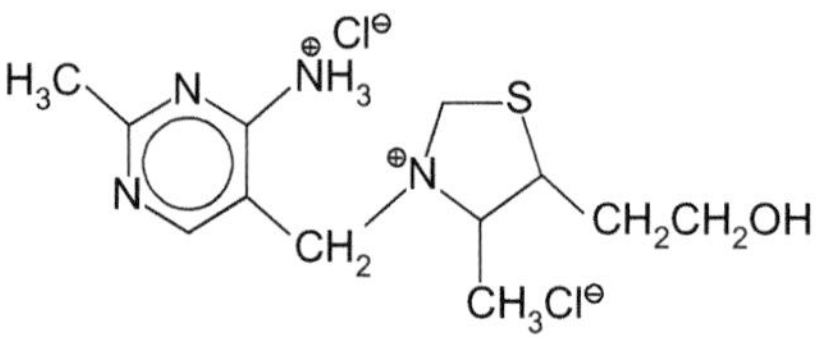

Thiamine, vitamin B, Anti-beriberi factor

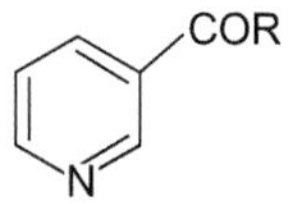

COR

Nicotinic acid, R = OH
Nicotamide, R = NH$_2$

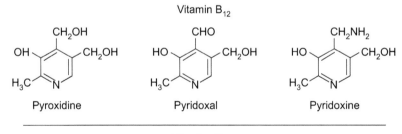

Vitamin B$_{12}$

Pyroxidine

Pyridoxal

Pyridoxine

Vitamine B$_6$

(±) Isopelletrine

Nicotine

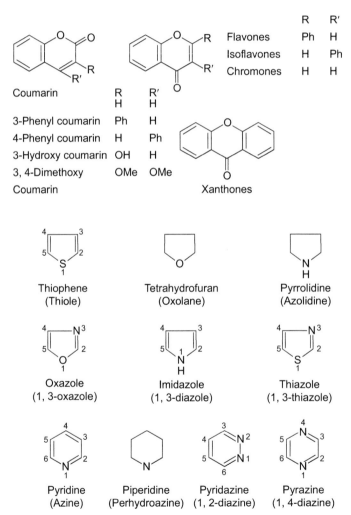

Morphine

In addition to the above types, another type of heterocyclic compounds contain oxygen as the hetero atom. These are mostly naturally occuring. Some examples include.

	R	R'
Flavones	Ph	H
Isoflavones	H	Ph
Chromones	H	H

Coumarin

	R	R'
	H	H
3-Phenyl coumarin	Ph	H
4-Phenyl coumarin	H	Ph
3-Hydroxy coumarin	OH	H
3, 4-Dimethoxy Coumarin	OMe	OMe

Xanthones

Thiophene
(Thiole)

Tetrahydrofuran
(Oxolane)

Pyrrolidine
(Azolidine)

Oxazole
(1, 3-oxazole)

Imidazole
(1, 3-diazole)

Thiazole
(1, 3-thiazole)

Pyridine
(Azine)

Piperidine
(Perhydroazine)

Pyridazine
(1, 2-diazine)

Pyrazine
(1, 4-diazine)

9.1.3 Nomenclature

Most of the heterocyclic compounds have been assigned trival names, which are difficult to replace by systematic names, which are adopted by IUPAC for nomenclature of monocyclic ring systems having one hetero atom. This system specifies the size of the ring, number, kind and position of the hetero atom(s) by providing an appropriate suffix and prefix as per a set rules described below.

1. The size of the ring is denoted by the stem, ir, et, ol, in, ep, oc, on or ee for 3⁻, 4⁻, 5⁻, 6⁻, 7⁻, 8⁻, 9⁻, and 10⁻ membered rings respectively.

2. The hetero atoms are designated for prefixes oxa (for oxygen), thia (for sulphur) and aza (for nitrogen). In case a ring contains two or more different hetero atoms, oxygen takes precedence over sulphur and sulphur over nitrogen. As an examples, a heterocycle containing one oxygen and one nitrogen atom is referred to as oxaza and thiaza denotes a heterocycle which contains one sulphur an one nitrogen atom.

3. The numbering of the ring begins with the hetero atoms and proceeds in a fashion so as to give lowest numbered positions to the substituents. In case the ring contains two or more different hetero atoms, oxygen takes precedence over sulphur and sulphur over nitrogen.

4. The degree of unsaturation in the ring is indicated by putting a suitable suffix. These suffixes change to some extent according to whether or not the ring contains nitrogen. The suffixes of some simple heterocyclic compounds in given in the table below.

Table 9.1: Suffixes of some simple heterocyclic compounds

Ring size	Stem	With nitrogen		Without nitrogen	
		Unsaturated	Saturated	Unsaturated	Saturated
3	–ir–	–irine	–iridine	–irene	–irane
4	–et–	–ete	–etidine	–ete	–etane
5	–ol–	–ole	–olidine	–ole	–olane
6	–in–	–ine	–inane	–in	–ane

The partially reduced heterocyclic compounds are referred to as dihydro or tetrahydro derivatives of the parent unsaturated compound. The preefix for a saturated nitrogen heterocyclic compound is perhydro. Some simple examples are given below with their trival names as well as IUPAC names (in parentheses). In case the name contains two vowels coming together, then 'a' of the prefix is omitted. (for example oxirane instead of oxairane).

| Ethylene oxido | Ethyleneimine | Furan | Pyrrole |
| (oxirane) | (Aziridine) | (oxole) | (Azole) |

9.2 THREE-MEMBERED HETEROCYCLIC COMPOUNDS

The saturated analugoues of three membered heterocyclic compounds are oxirane (epoxide), aziridine and thiirane.

9.2.1 Oxiranes (Epoxides)

9.2.1.1 Preparation

It is obtained commercially by catalytic oxidation of ethylene by air.

These can also be obtained by epoxidation of alkenes with peracids, most commonly with m-chloroperbenzoic acid.

Conjugated alkenes which have electron withdrawing groups, such as a carbonyl group, do not react with peracids. However, such alkenes can be epoxidised with alkaline hydrogen peroxide.

Alkenes can also be converted into epoxides by treatment with hypochlorous acid, followed by cyclisation of the formed halohydrin with base.

Epoxidation of alkenes can also be effected by treatment with dimethyl dioxirane which is generated in situ by the reaction of oxone with acetone.

9.2.1.2 Properties

The oxiranes are highly strained molecules. The ring opening releases the strain. The acid catalysed ring opening of epoxides gives 1, 2-diols (glycols). This is an S_N^2 reaction in which water acts as a nucleophile, which attacks the carbon on the side opposite to oxygen resulting in inversion of configuration at one of the carbon atoms. Epoxidation of alkene followed by acid catalysed hydrolysis is a convensent method for hydroxylation of alkene.

In case of an unsymmetrical epoxode, like isobutylene oxide. The two carbon atoms (of the epoxide ring) are not equivalent and so may give two products. It has been established that the preferred point of attack by the nucleophile (e.g., water or alcohl) depends on whether the reaction is catalysed by an acid or a base. In general, in acidic conditions, the nucleophile attacks the more substituted carbon atom and under alkalene conditions, the nucleophile attacks the less substituted carbon atom.

Acid catalysed ring opening

Isobutylene oxide

In this case, the bond indicated by dotted line is weak and the bond breaks.

T.S

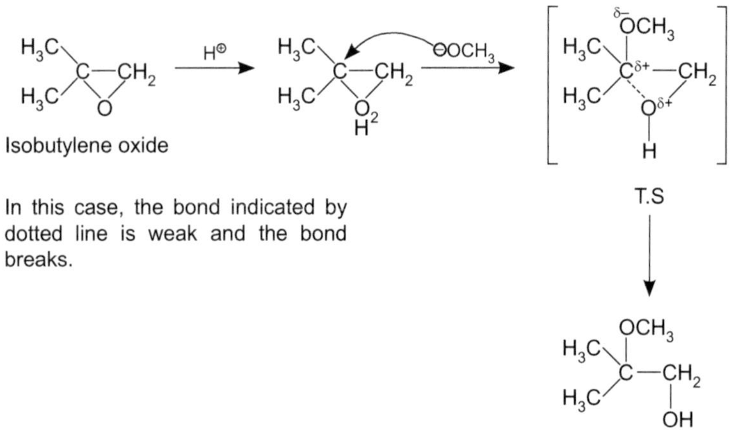

Base catalysed ring opening

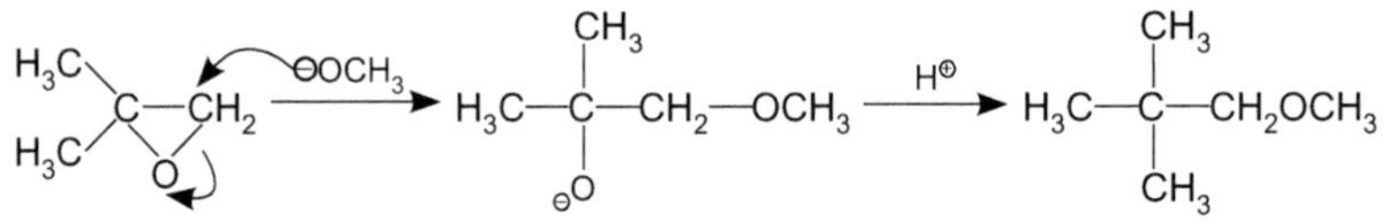

In the above case, the nucleophile attacks the least substituted carbon of the epoxide. In this case the controlling factor is sceric hinderance.

The epoxode ring can also be opened up by reaction with grignard reagent. This is a convenient method for the synthesis of primary alcohol containing two additional carbon atoms.

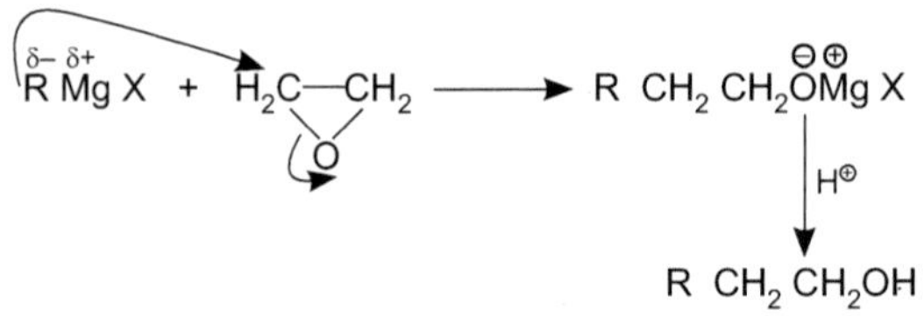

Epoxides undergo deoxygenation on heating with *n*-butyl phusphine to give olifins having opposite configuration to that of the starting epoxide.

9.2.3 Aziridines

9.2.3.1 Synthesis

Aziridines are synthesised from appropriately substituted alkenes by following methods.

(*i*) Insersion of nitrenes into double bond of an alkene. The nitrenes are generated by the photolysis or thermolysis of azides. These nitrenes get inserted into the double bond of an alkene to give azindines

(*ii*) Alkenes can be converted into aziridines by a three step process involving chloronitrosation, reduction and final cyclisation with base.

(*iii*) β-Amino alcohols (obtained by treatment of epoxides with ammonia) on acid-catalysed cyclisation give aziridines.

9.2.3.2 Properties

Suitably substituted aziridines (which have substituents on the trivalent nitrogen occupying a different plane than that at the ring) can in principle be resolved into optically active enantiomers.

However, nitrogen undergoes rapid inversion of configuration at room temperature. It is however, possible to to effect resolution at very low temperature.

Aziridines undergo ring opening by treatment with HCl or $C_2H_5NH_2$ with inversion of configuration at the point of attack.

Aziridinium ion (unstable)

Optically active   Meso

Aziridines condense with aldehydes to yield oxazolidines.

Aziridines undergo Friedel—Crafts reaction with benzene to give β-phenylethylamine.

β-Phenylethylamine

The H atom attached to nitrogen in aziridine is reactive and can be substituted.

Azinidines can be converted into alkenes by treatment with nitrosyl chloride.

9.2.4 Thiiranes

Also known as episulphides, the thiirazanes have the lowest strain energy of the three membered hyeterocycles.

9.2.4.1 Synthesis

(*i*) The reaction of epoxides with thiourea or throcyanate salts give thiiranes.

(*ii*) Thiiranes can also be obtained by intramolecular cyclisation of 2-halomercaptans with a base.

9.2.4.2 *Properties*

Thiiranes are less reactive towards electrophilic reagents. This is due to lower electron density at the sulphur atom than at oxygen atom in oxiranes.

The reaction of thiiranes with HCl results in ring opening of thiiranes giving the formation of halomercaptans. the ring opening occurs stereospecifically with inverson at the point of attack.

Ring opening of thiirenes can also be effected by treating with $LiAlH_4$, alcohol, amines and acid chlorides.

On treatment of thiirane with tertiary phosphines there is stereospecific removal of sulphur resulting in the formation of olefin having original configuration of thiirane.

9.3 FOUR MEMBERED HETEROCYCLIC COMPOUNDS

Four membered heterocyclic compounds are considered to be derived from cyclobutane by replacement of a CH_2 group by a hetero atom (O, NH and S). These are oxetanes, azetidines and thretanes.

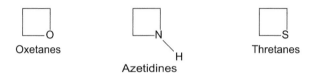

Oxetanes   Azetidines   Thretanes

9.3.1 Oxetanes

Also called oxycyclobutanes, the oxetane molecule is not a perfect square, since C—C bond length (1.54 Å) is greater than C—O bond length (1.46 Å).

9.3.1.1 Synthesis

(*i*) Intramolecular cyclisation of 1, 3-halohydrins in presence of a base. The rate of cyclisation can be increased by acetylation of the halohydrin followed by cyclisation.

$$H_2C\!-\!CH_2\!-\!Br$$
$$H_2C\!-\!O\!-\!H$$

1, 3-Halohydrin

$$\xrightarrow{\text{base}}$$

$$H_2C\!-\!CH_2$$
$$H_2C\!-\!O$$

acetylation

$$H_2C\!-\!CH_2\!-\!Br$$
$$H_2C\!-\!O\!-\!COCH_3$$

$$\xrightarrow{\text{base}}$$

(*ii*) From oxiranes by treatment with sulphur ylids. The reaction involves methylene insersion from the less hindered side.

$$Ph\!-\!HC\!-\!CH_2$$
$$\diagdown O \diagup$$

Oxirane

$$\longrightarrow$$

$$Ph\!-\!HC\!-\!CH_2$$
$$O\!-\!CH_2$$

Oxetanes

$$\left[(H_3C)_2 \overset{\|}{S}\!=\!CH_2 \quad\longleftrightarrow\quad (CH_3)_2\, \overset{+}{\underset{\|}{S}}\!-\!\overset{-}{C}H_2 \right]$$
$$\qquad\quad O \qquad\qquad\qquad\qquad O$$

Sulphur ylide

(*iii*) **Paterno-Buchi Reaction:** It involves photochemical cycloaddition reaction of carbonyl compounds with olefins.

9.3.1.2 Properties

The most important reactions of oxetanes is the ring opening which takes place in presence of acids.

However, in case of unsymmetrical oxetanes, two products are obtained.

In case of 2-phenyl oxetane only one product is obtained as the carbocation intermediate is stable and so only one product is obtained.

Friedel-crafts reaction of 2-methyl or 2-phenyl oxetane also give a single product due to the generation of stabilized carbocation.

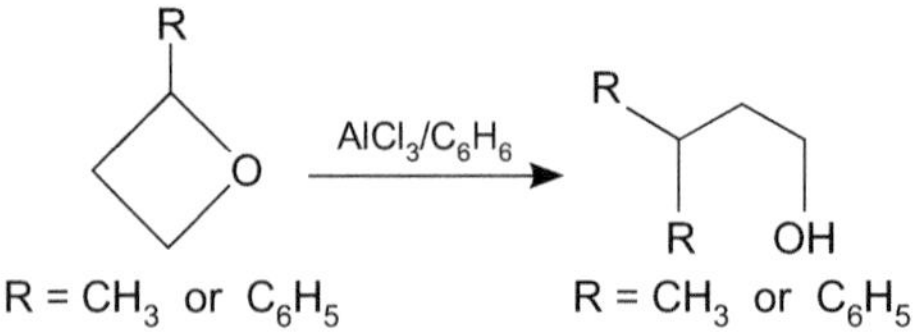
R
$AlCl_3/C_6H_6$
R
R OH
$R = CH_3$ or C_6H_5
$R = CH_3$ or C_6H_5

Oxetanes also undergo ring opening with nucleophiles like CH_3NH_2, $C_6H_5CH_2SH$. But in this case higher temperature is required.

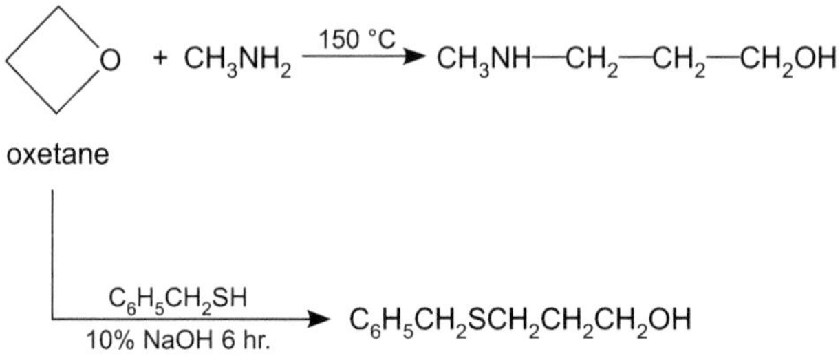
+ CH_3NH_2
150 °C
$CH_3NH—CH_2—CH_2—CH_2OH$
oxetane
$C_6H_5CH_2SH$
10% NaOH 6 hr.
$C_6H_5CH_2SCH_2CH_2CH_2OH$

9.3.2 Azetidines

9.3.2.1 Synthesis

(*i*) Intramolecular cyclisation of γ-haloalkyl amines in presence of base.

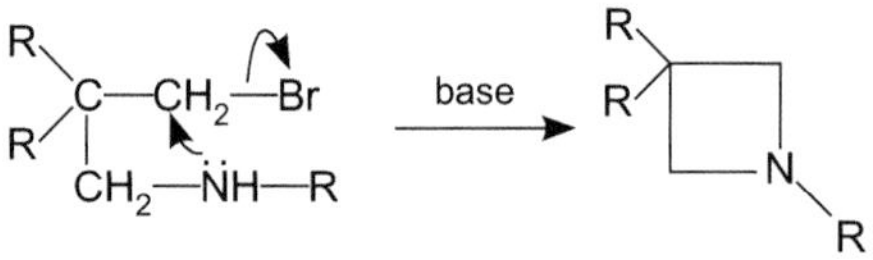
base

This method is useful for the synthesis of 3-substituted azitidines.

(*ii*) Azetidine itself can be prepared by cycloaddition of 3-chloropropyl-bromide with p-toluenesulfonamide followed by reduction of the formed N-substituted azetidine.

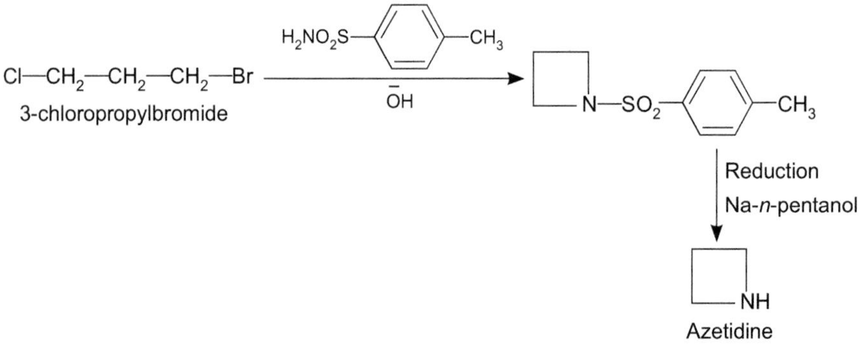
$Cl—CH_2—CH_2—CH_2—Br$
3-chloropropylbromide
$\bar{O}H$
Reduction
Na-*n*-pentanol
NH
Azetidine

9.3.2.2 Properties

Azetidines are quite resistant to the action of bases and nucleophiles. However, azetidines undergo ring opening by H_2O_2 or HCl.

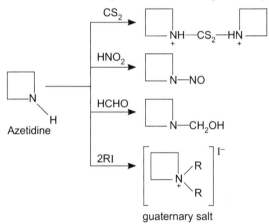

As secondary amines, azetidine reacts with CS_2, HNO_2 and HCHO. However, as tertiary amine it reacts with alkyl halide to form quaternary salt.

9.3.3 Thietane

9.3.3.1 Synthesis

(*i*) 1-Bromo-3-chloropropane on treatment with one equivalent of thiourea gives the corresponding monothrouronium salt, which on heating with NaOH solution gives thietane.

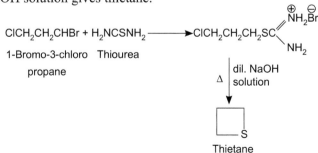

(*ii*) 1, 3-Dibromopropane on heating with aqueous Na_2S in presence of a PTC gives thietane

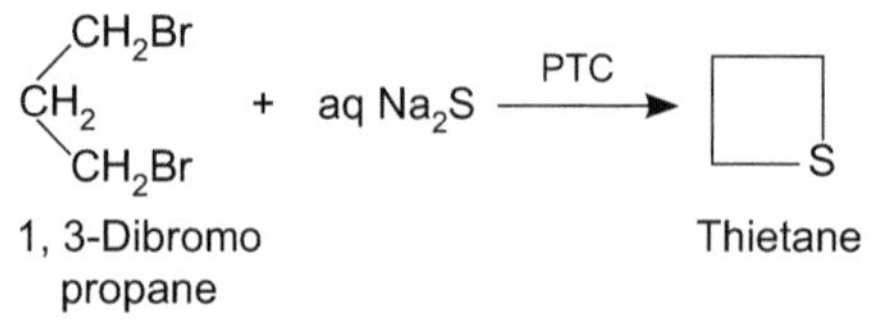

1, 3-Dibromo
propane

Thietane

The PTC used in the above reaction is hexadecyl triethyl ammonium chloride.

9.3.3.2 Properties

On heating in gaseous phase, thietane undergoes ring opening to give ethylene and throformaldehyde.

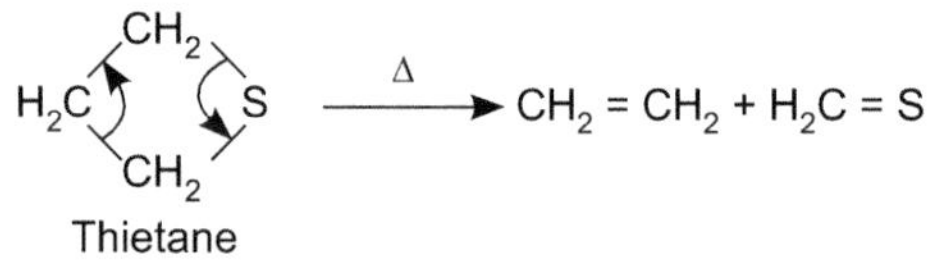

Thietane

Ring opening of thietane can also be effected by treatment with chlorine at −70 °C.

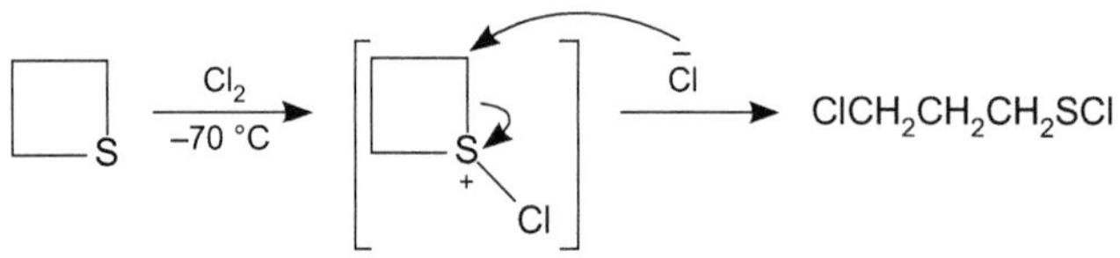

Thietane on oxidation with H_2O_2 give the corresponding sulfoxide

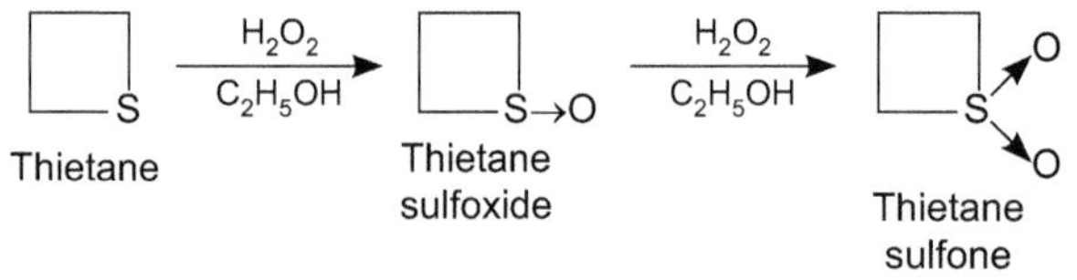

Thietane   Thietane
                       sulfoxide

Thietane
sulfone

9.4 FIVE MEMBERED HETEROCYCLIC COMPOUNDS

9.4.1 Introduction

The simple five membered heterocyclic compounds containing one hetero atom are furan, pyrrole and thiophen

| Furan | Pyrrole | Thiophen |

As seen from the structures of the three hydrocyclic compounds, it is expected that they should have properties characteristic of a conjugated diene and of an ether, amine or a sulphide in case of furan, pyrrole, thiophen respectively. It has, however been found that these heterocycles do not exhibit any such properties. In fact, they possess considerable aromatic character. This has been shown by the fact that these heterocyclic compunds readity undergo electrophilic substitution reactions like nitration, sulphonation, halogenation and Friedel-crafts reaction. Their heats of combustion indicate resonance stabilization to the extent of 16 Kcal/mole for furan, 21 Kcal/mol for pyrrole and 29 Kcal/mole for thiophen, but is larger in magnitude than that of conjugated dienes (about 4 Kcal/mole). On the basis of the above it is clear that the above three heterocyclic compounds possess aromatic character.

The aromatic character of these heterocyclic compounds arise from the delocalization of four carbon π electrons and two paireal electrons denoted by the hetero atom (O, N and S), thus forming a sextect of electrons which is characteristic of aromatic systems.

The aromatic character of these three heterocyclic compounds is also well understood by consideration of their molecular orbitals. All these heterocycles are planar pentagans comprising of sp^2-hydridized carbon atoms. This arrangement leaves each of the four carbon atoms with a p-orbital which overlaps with a doubly filled p-orbital of the hetero atom. This results in the formation of a π-cloud above and below the plane of the ring. As there are six electrons in these systems, they are aromatic in character. The aromatic character of these heterocyclic compounds also finds support in the observation in all these heterocyclic compounds is that the bond lengths of C—C, C—N, C—S and C—O bonds are intermediate in length between the normal and double bond lengths.

The structures of the three heterocycles, furan, pyrrole and theophen is considered to be as resonance hybrid of the following conaconical structures.

X= O, NH or S

As seen, the ring carbons in these heterocycles acquire some what negative charge. In all these cases, electrophilic substitution occurs as position 2. This is well understood by the fact that the attachment of the electrophile at position 2 affords most stable intermediate carbocation compared to the attachment of the electrophile at position 3. This is shown by the following for furan.

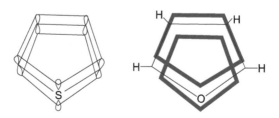

(Three canonical structures)

(Two canonical structures)

As seen, attack at C_2 produces more canonical structures compared to 2 cononical structures by attack at C_3. So substitution at C_2 is preferred.

On the basis of reactions of furan, pyrrole and thiophen it been established that furan is more reactive than pyrrole and thiophen is the least reactive.

Furan > pyrrole > thiophen

An example illustrating that furan is more reactive than thiophen is given below.

9.4.2 Furan

The importance of furan is due to its derivatives, which are of industrial importance. These derivatives include furfural (which is used for the preparation of dyes, plastics) and THF (which is used as a industrial solvent).

Molecular orbital picture of Furan

The furan has a planar pentagonal structure; the four carbon atoms and oxygen atom are sp^2 hybridized. The sp^2-hybridized orbitals overlap with each other and also with the s-orbital of hydrogen. The two lone pairs of electron on oxygen are in different π orbital. The molecular orbital picture of furan is represented as shown on page 110.

9.4.2.1 Synthesis

(*i*) Furan is obtained from pentoses (which are found in oat hulls) by treatment with hydrochloric acid. Various steps involved are given below.

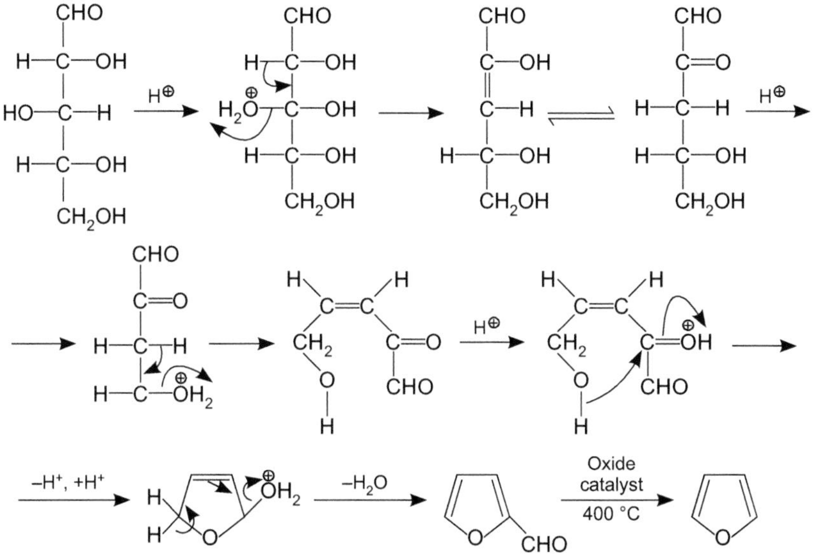

Furtural   Furan

(*ii*) **Paal-Knorr synthesis:** It involves cyclisation of 1, 4-dicarbonyl compounds with reagents like H_2SO_4, $ZnCl_2$, P_2O_5 etc.

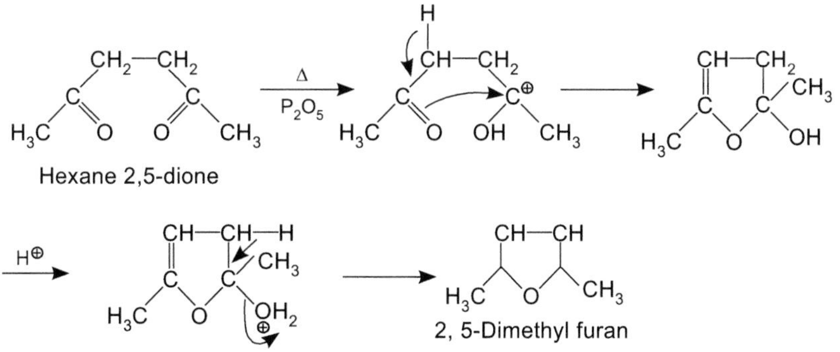

Hexane 2,5-dione

2, 5-Dimethyl furan

This is a general method for the synthesis of furan derivatives.

(*iii*) **First-Benary Synthesis**

The reaction of β-ketoester with an α-haloketone gives furan derivatives.

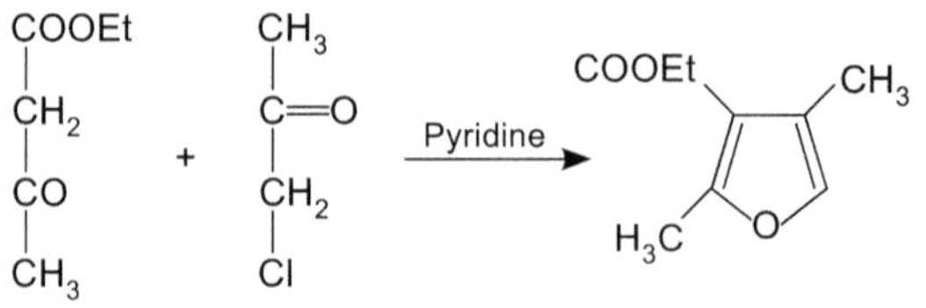

Acetylene dicarboxylic esters have also been used for the synthesis of furan derivatives.

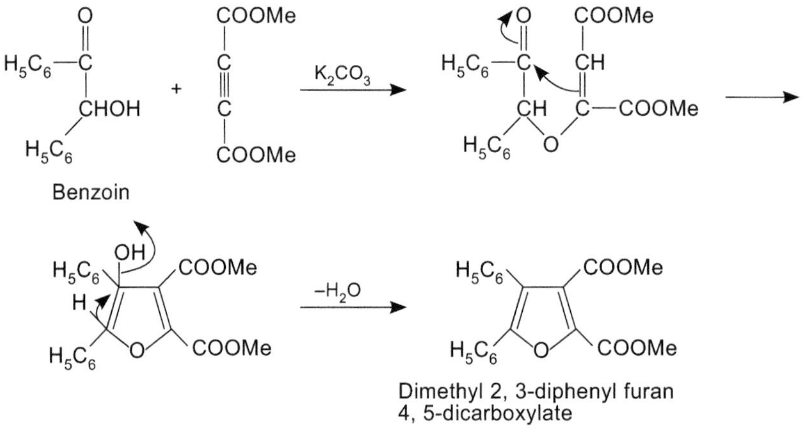

Benzoin

Dimethyl 2, 3-diphenyl furan
4, 5-dicarboxylate

(*iv*) The reaction of monosodium salt of ethyl acctoacetate with iodine gives diaceto succinic ester, which on cyclisation with H_2SO_4 yields 2, 5-dimethylfuran-3, 4-dicarboxylic acid.

$$CH_3COCH_2COOEt \xrightarrow{NaOEt} CH_3CO\overset{-}{C}HCOOEt \xrightarrow[-2\ NaCl]{I_2}$$

Ethylacetoacetate

Na+
monosodium salt

$$CH_3COCHCOOEt$$
$$\ \ \ \ |$$
$$CH_3COCHCOOEt \xrightarrow[H_2SO_4]{\Delta}$$

Diacetosuccinic
ester

2, 5-Dimethyl furan
3, 4-dicarboxylic acid

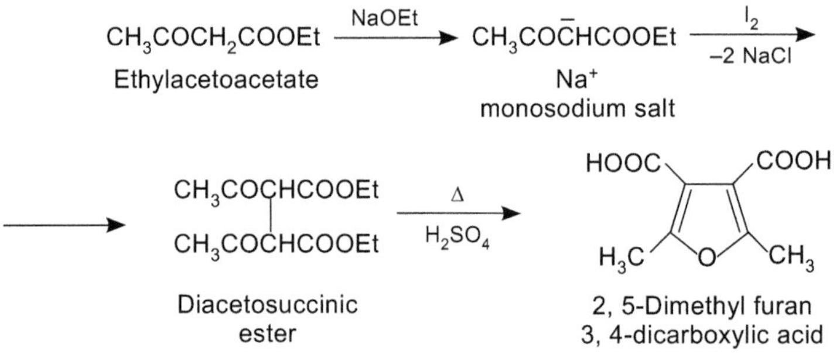

9.4.2.2 Properties

Furan is a colourless liquid, b.p. 32 °C. It undergoes following reactions.

(*i*) **Electrophilic substitution:** As already stated furan behaves as an aromatic compound and undergoes electrophilic substitution reactions preferentially at 2-position.

Furan gets polymensation under acidic conditions, so appropriate reagents and reaction conditions must be used.

As an example, nitration of furan is performed with acetyl nitrate and not the usual nitrating mixture which leads to polymerisation. Similarly furan 2-sulphonic acid is prepared by using pyridine-sulphur trioxide.

Also, halogenation of furan with halogen librates halogen acids, which also cause polymerisation of furan, so halogenated furans are obtained by an indirect method involving halogenation of furoic acid followed by decarboxylation of the formed product.

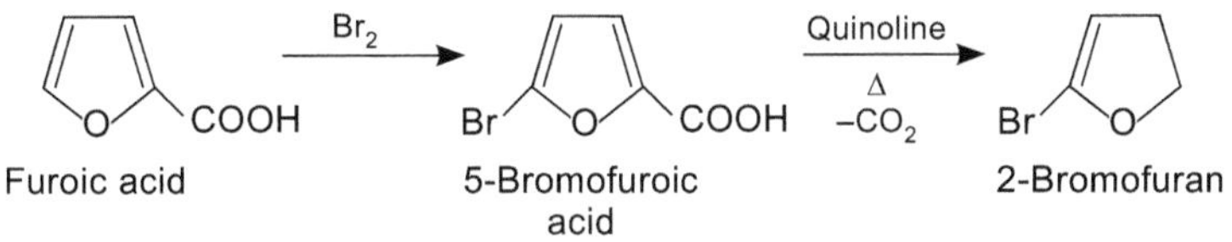

Furoic acid   5-Bromofuroic acid   2-Bromofuran

2-chloro or iodofuran can be obtained from 2-chloromercurifuran by treatment with Br_2 or I_2. 2 Acetyl furans can be similarity obtained from 2-chloromercurifuran with acetyl chloride.

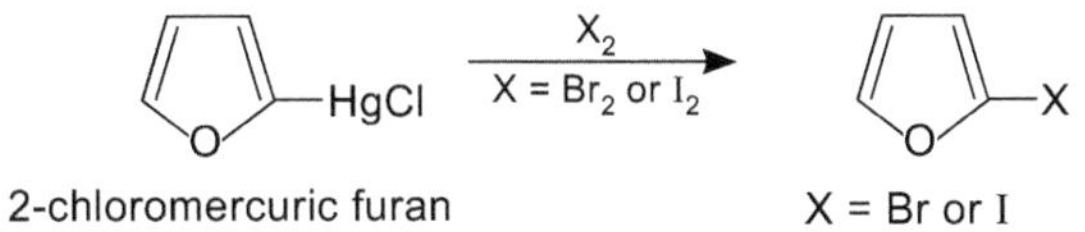

2-chloromercuric furan   X = Br or I

Infact, 2-chloromercurifuran is a useful synthetic intermediate for the synthesis of 2-substituted furans, since the mercuri-group can be easily replaced by other groups.

Furoic acid is obtained by treatment of furan with *n*-butyllithium, followed by carboxylation of the organolithium compound.

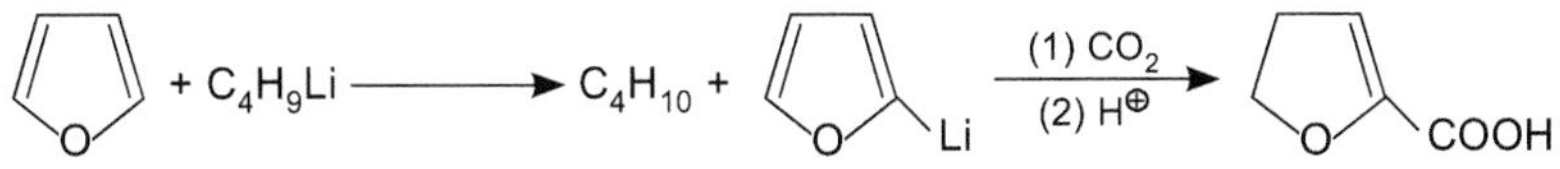

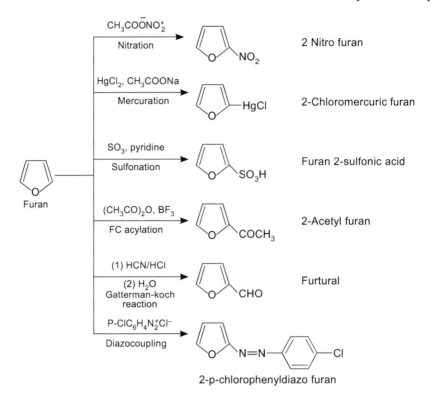

(*ii*) **Reduction:** Catalytic reduction of furan yields tetrahydrofuran (THF), an industrially important solvent. However, reduction with Pt in presence of acetic acid gives *n*-butyl alcohol. Reduction with Na/ alcohol or Zn/HOAc can be controlled to give 1, 2 or 1, 1-addition.

(*iii*) **Diels-Alder Reaction:** Furan being least aromatic of the five-membered heterocycles (due to greater electronegativity of oxygen and so is reluctant to part with the pair of electrons for delocalization) behaves like a diene and undergoes Diels-Alder reaction.

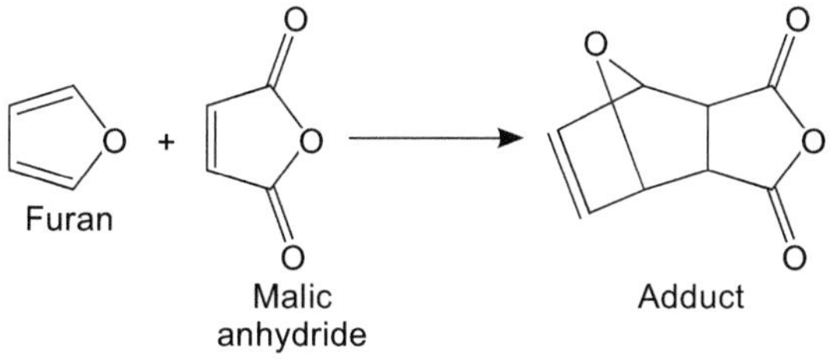

Furan

Malic
anhydride

Adduct

Diels-Alder reaction with maleimide at room temperature gives the endo adduct, which on heating at 90 °C rearranges to the exo adduct.

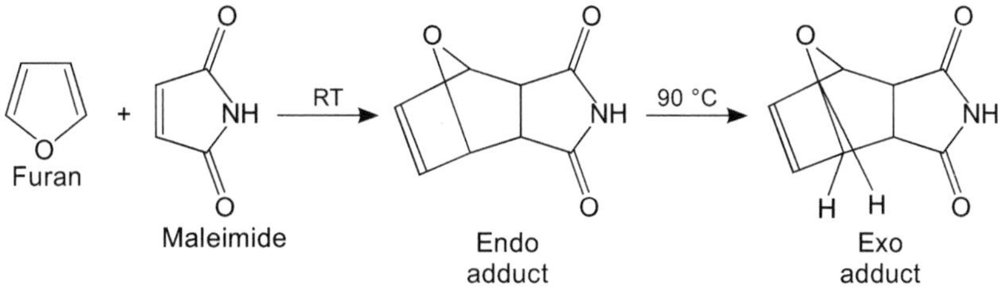

Furan

Maleimide

Endo
adduct

Exo
adduct

(*iv*) **Reaction with carbenes:** Furan undergoes cycloaddition reactions with carbenes (generated in situ by a number of methods) to yield cyclopropane derivatives. Cycloaddition with ethoxycarbonylnitrene yields aziridine derivatives.

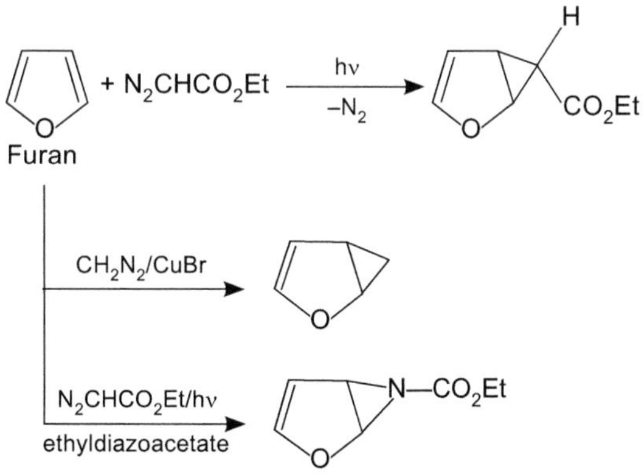

Furan

ethyldiazoacetate

(*v*) **Photochemical cycloaddition:** Furan undergoes photochemical cycloaddition reaction with carbonyl compounds to yield oxetanes.

Furan   Propinaldehyde

oxetane derivative

A number of derivatives of furan are of industrial importance. A common derivative is furaldehyde (furfural), which behaves like benzaldehyde and undergoes cannizzaro reaction, benzoin condensation, perkin reaction and classen-smidt reaction.

Furtural

Furfuryl alcohol

Furoic acid (sodium salt)

Furoin

Furylidene acetone

9.4.3 Pyrrole

Pyrrole is known to occur in coal bar and bone oil, from which it is obtained on a commercial scale.

As in the case of furan, the structure of pyrrole can be described as a hybrid of several structures (section 1.8.6.1). It has a planar pentagonal frame work, consisting of sp^2 hybridized carbon atoms. This arrangement leaves with a p-orbital of four carbon atoms which overlap with a doubly filled p-orbital of the hetero atom resulting in the formation of a π cloud above and below the plane of the ring. Thus, the orbital of pyrrole is represented as shown.

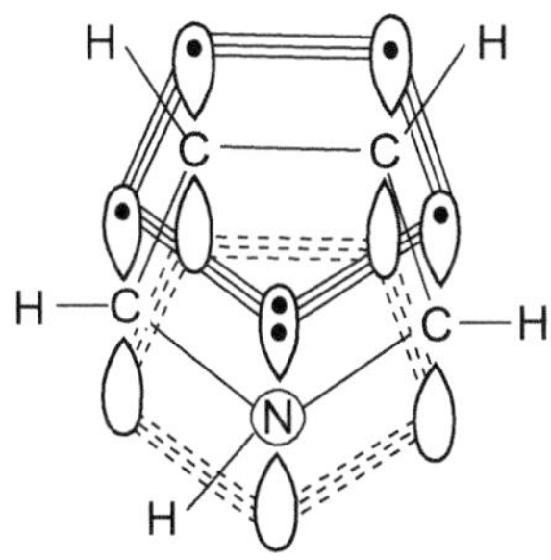
H
H
C
C
H—C
C—H
N
H

Orbital pictures of pyrrole

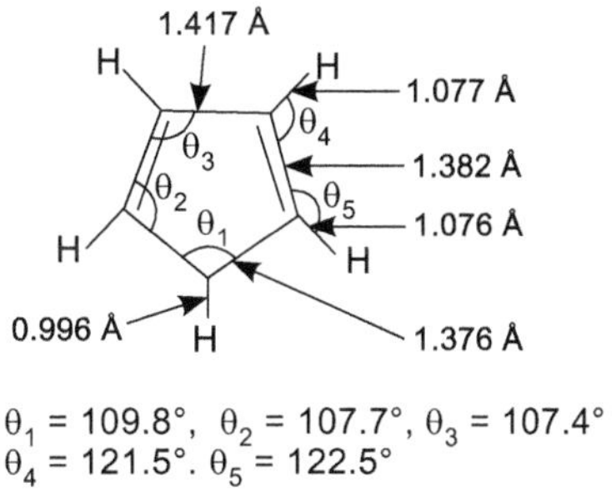
1.417 Å
1.077 Å
1.382 Å
1.076 Å
0.996 Å
1.376 Å
θ1 = 109.8°, θ2 = 107.7°, θ3 = 107.4°
θ4 = 121.5°. θ5 = 122.5°

Bond angles and bond lengths of pyrrole

The structure of pyrrole is supported on the basis of the measurements of bond lengths, dipole moments, UV spectra, microwave spectra and heats of combustion. The bond length in pyrrole are between normal and double bond lengths of C—C and C—N bonds.

9.4.3.1 Synthesis

(*i*) ***Paal-Knor synthesis:*** It involves the reaction of an enolizable 1, 4-diketone with ammonia or a primary amine.

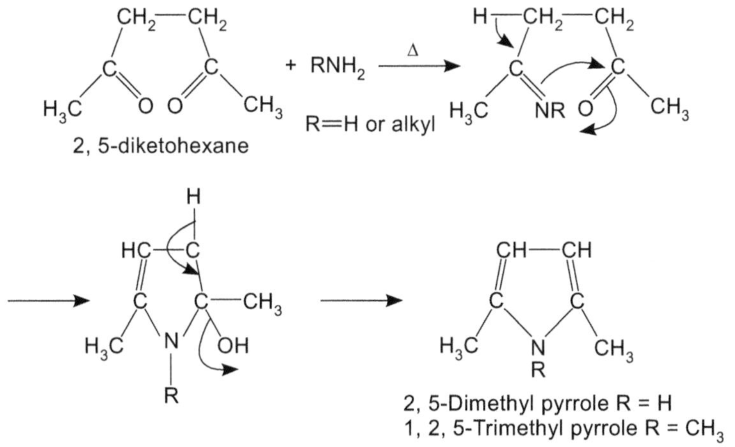
+ RNH2
Δ
R=H or alkyl
2, 5-diketohexane
2, 5-Dimethyl pyrrole R = H
1, 2, 5-Trimethyl pyrrole R = CH3

(*ii*) ***Hentzch Pyrrole Synthesis:*** It involves conensation of β-keto ester with an α-haloketone in presence of ammonia or a primary amine. Various steps involved are.

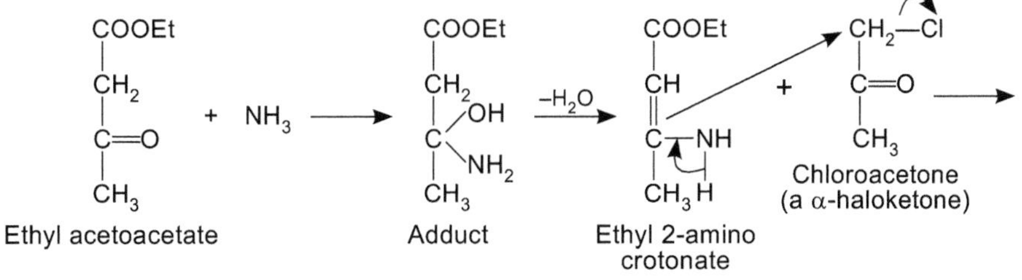
+ NH3
–H2O
Ethyl acetoacetate
Adduct
Ethyl 2-amino
crotonate
Chloroacetone
(a α-haloketone)

Ethyl 2, 5-dimethyl
pyrrole 4-carboxylate

(iii) **Knorr Pyrrole Synthesis:** A widely used synthesis for synthesis of a
variety of pyrroles, involves condensation of an α-aminoketone with α
β-keto ester in presence of acetic acid.

Diethyl 2, 4-dimethyl
pyrrole 3, 5-dicarboxylate

(iv) **From Bone Oil:** As already mentioned, pyrrole is obtained on a large
scale from bone oil. Various steps involved are given below.

(v) **From Furan:** Pyrrole is manufactured from furan by passing a mixture
of furan, ammonia and steam over alumina catalyst.

9.4.3.2 *Properties*

Pyrrole is a liquid, bp 131 °C. It is a weak base (*pKa* = 0.4) as the lone pair of electrons on nitrogen (which are responsible for the basicity of amines) is not available due to its delocalization with the ring completing the aromatic sextet.

It undergoes following reactions.

(*i*) **Electrophilic substitution:** As already stated, electrophilic substitution in pyrrole take place at 2-position. Some of the typical electrophilic substitution reactions of pyrrole are given below.

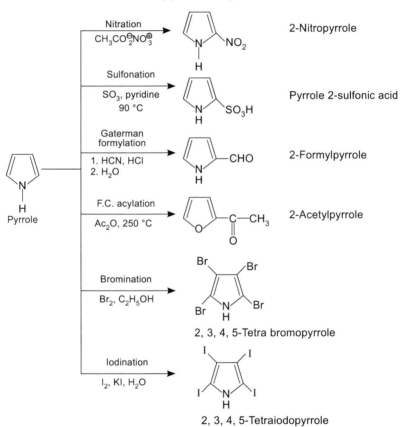

(*ii*) **Nucleophilic Substitutions:** Nucleophilic substitutions are not normally encountered in pyrrole. This is attributed to excessive π-electron character due to which nucleophilic substitution are not normally possible. However, if pyrrole is protonated or substituted with electron withdrawing substitutions, nucleophilic substitution may occur. An example is given below.

3, 4-Dinitro
pyrrole

(*iii*) **Acidic Character:** Pyrrole on heating with solid KOH, the imino hydrogen is replaced with potassium indicating the acidic character of pyrrole.

(*iv*) **Reduction:** Reduction of pyrrole will zinc and acetic acid gives pyrroline, which on heating with hydrodic acid and red phosphorous given pyrrolidine (tetra hydro pyrrole). Tetra hydropyrrole can also be obtained from pyrrole by calalytic reduction.

In pyrrolidine, the aromatic character of pyrrole is destroyed. So pyrrolidine shows typical properties of aliphatic secondary amine. In this case, the extra pair of electrons on N is available for sharing with a proton.

(*v*) **Reaction with carbenes:** Unlike furan, pyrrole yields the normal product of electrophilic substitution in presence of a carbene precursor.

Pyrrole reacts with dichlorocarbene, (generated in situ by the reaction of $CHCl_3$—NaOH) gives pyrrole-2-carbaldehyde along with 3-chloropyridine. The lather product is obtained by ring expansion involving insersion of dichloro carbene into C_2—C_3 bond of pyrrole.

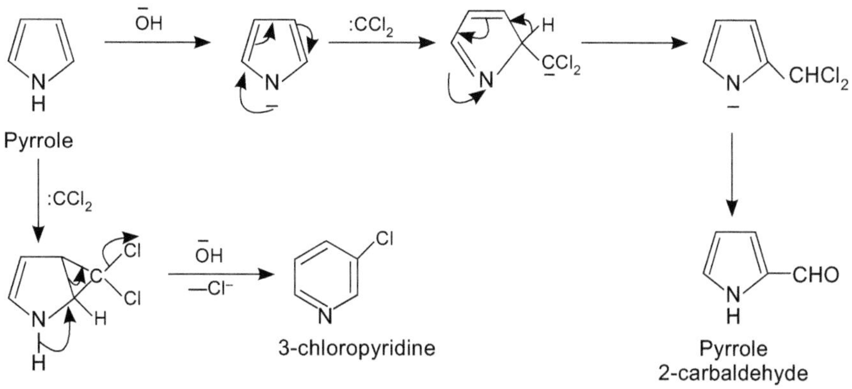

Pyrrole

3-chloropyridine

Pyrrole 2-carbaldehyde

It is interesting to note that pyrrole-2-aldehyde unlike furturaldehyde does not undergo cannizzores reaction or benzoin condensation.

(*vi*) **Reaction with nitrenes:** The reaction of pyrrole with nitrenes (generated by copper catalysed decomposition of ethyl azidoformate) gives 2-amino-1-ethoxycarbonyl pyrrole. The reaction involves insersion of nitrene into C_2—C_3 bond followed by rearrangement of the formed adduct.

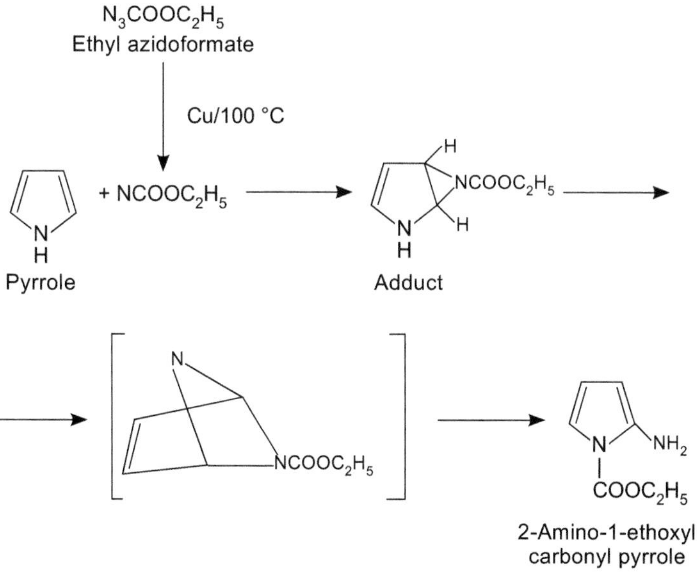

$N_3COOC_2H_5$
Ethyl azidoformate

Cu/100 °C

Pyrrole   + $NCOOC_2H_5$ Adduct

2-Amino-1-ethoxyl carbonyl pyrrole

(*vii*) **Reaction with benzyne:** The reaction of pyrrole with benzyne involves a Michael-type condensation giving 2 phenyl pyrrole.

(*viii*) **Diel-Alder reaction:** Unlike furan, pyrrole does not undergo Diel-Alder reaction on treatment with maleic anhydride. In this case the product obtained is a substitution product.

Pyrroles having electron withdrawing groups on N behave as diene in Diels-Alder reaction. In this case, due to arromatic character of pyrrole only Michael addition product is obtained.

9.4.4 Thiophen

Thiophen is found to occur in coal tar in small amounts and is invariably obtained in the benzene fraction of coal tar. As the b.p of thiophen (84 °C) is close to the B.p of benzene (80 °C), it is not possible to separate the two by fractional distillation. However, the separation can be effected as thiophene undergoes sulphonation more readity than benzene. Thus, the two components can be separated by shaking the mixture with cold H_2SO_4, and the water soluble thiophen sulphonic acid is washed out with water from water insoluble benzene layer. The thiophen-2-sulphonic acid in the aqueous phase is refluxed with mercuric acetate. Thiophene gets mercurated. The mercurated thiophen on distillation with HCl gives thiophen, which is subsequently repurified by fraditional distillation.

Thiophen is an aromatic five numbered heterocyclic containing sulphur. Compared to furan or pyrrole, thiophen is less reactive. It is virtually non-basic and does not form salts. Its molecular orbital picture is shown below.

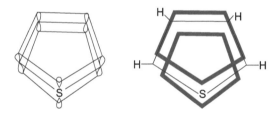

Molecular orbital picture of Thiophen

As in the case of furan and pyrrole, in thiophen also, the contribution of two electrons by sulphur completes the aromatic sextet. Thiophen is more aromatic than other 5-membered heterocyclic compounds.

9.4.4.1 Synthesis

(*i*) Thiophen is manufactured by the cyclization of *n*-butane with sulphur.

$$CH_2\text{—}CH_2 \quad | \quad | \quad + \quad 4S \quad \xrightarrow{600\ °C} \quad \text{(thiophene)} \quad + \quad 3H_2S$$
$$CH_3 \quad CH_3$$

(*ii*) On a commercial scale, thiophen can also be obtained by passing a mixture of acetylene and H_2S over heated alumina.

$$2C_2H_2 + H_2S \quad \xrightarrow[400\ °C]{alumina} \quad \text{(thiophene)} \quad + \quad H_2$$

(*iii*) Thiophen is prepared in the laboratory by heating a mixture of sodium succinate and P_2S_3.

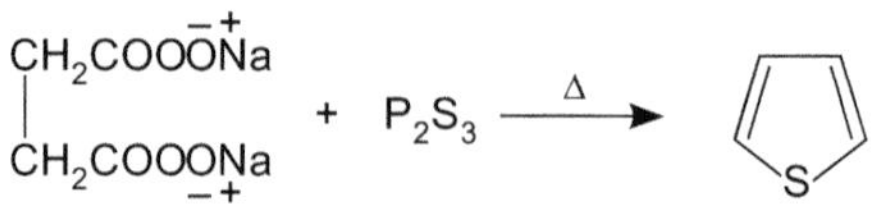

(*iv*) **Paal-Knorr Synthesis:** The method involves heating 1, 4-dicarbonyl compound with P_2S_5.

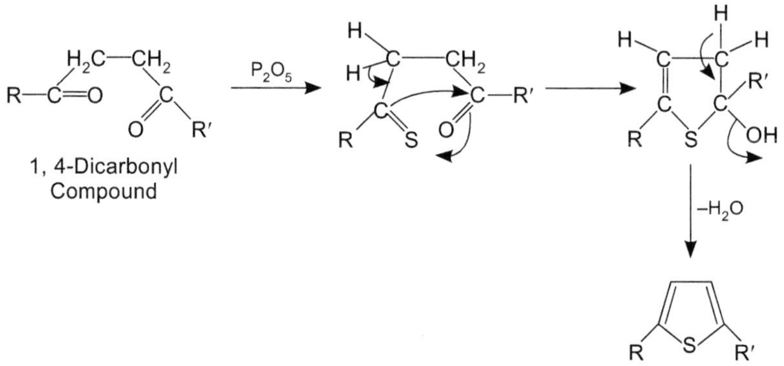

1, 4-Dicarbonyl
Compound

(*v*) Thiophen derivatives can be obtained by the reaction of α-diketones with diethyl thiodiacetate.

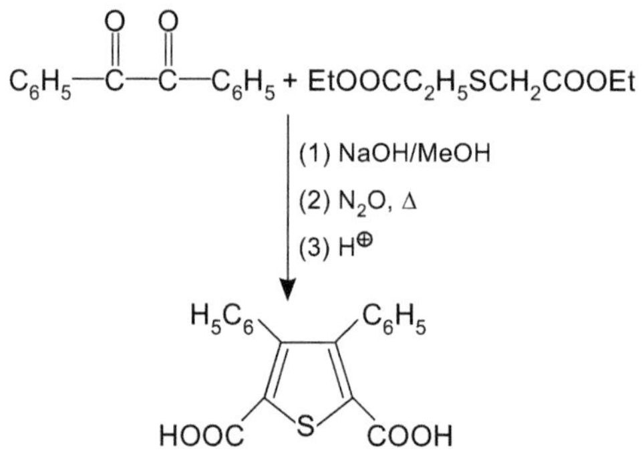

C_6H_5—$\overset{\overset{O}{\|}}{C}$—$\overset{\overset{O}{\|}}{C}$—$C_6H_5$ + $EtOOCC_2H_5SCH_2COOEt$

(1) NaOH/MeOH
(2) N_2O, Δ
(3) $H^\oplus$

9.4.4.2 Properties

As already stated, thiophen is less reactive than furan and pyrrole. Also, thiophen does not show basic properties. Following are given some of the typical reactions of thiophen.

(*i*) **Electrophilic substitution:** As in the case of furan and pyrrole electrophilic substitution take place at position 2. As thiophen is less reactive than furan or pyrrole, nitration or sultonation can be carried out with less problems. Some typical electrophilic substitution reactions of thiophen are:

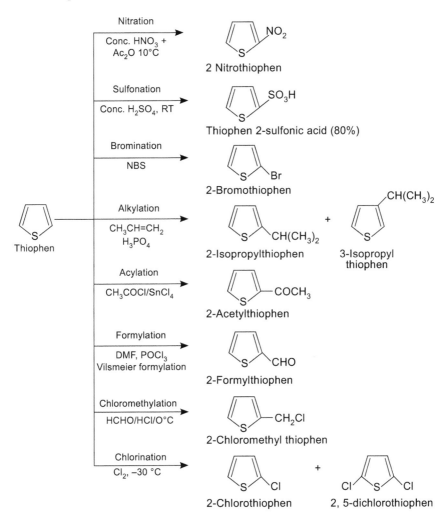

Nitration

Conc. HNO_3 + Ac_2O 10°C

2 Nitrothiophen

Sulfonation

Conc. H_2SO_4, RT

Thiophen 2-sulfonic acid (80%)

Bromination

NBS

2-Bromothiophen

Alkylation

$CH_3CH=CH_2$ H_3PO_4

2-Isopropylthiophen   3-Isopropyl thiophen

Acylation

$CH_3COCl/SnCl_4$

2-Acetylthiophen

Formylation

DMF, $POCl_3$ Vilsmeier formylation

2-Formylthiophen

Chloromethylation

HCHO/HCl/0°C

2-Chloromethyl thiophen

Chlorination

Cl_2, –30 °C

2-Chlorothiophen   2, 5-dichlorothiophen

Thiophen

(*ii*) **Nucleophilic substitution:** Nucleophilic substitution in thiophenes do not follow the usual course involving substitution on carbon bearing the leaving group, but involve introduction of nucleophile on carbon atom adjacent to the carbon atom having the leaving group; this is called **cine-substitution**. Some examples are

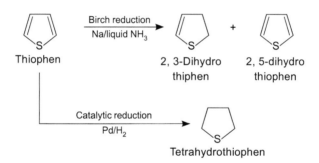

(*iii*) **Reduction:** Reduction of thiophen with sodium in liquid ammonia (**Birch reduction**) gives 2, 3- and 2, 5-dihydrothiophenes. However, catalytic reduction gives tetrahydrothiophen.

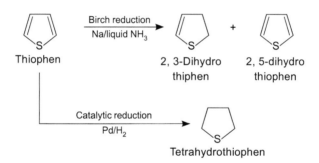

(*iv*) **Reaction with carbenes:** Thiophen reacts with carbenes (generated in situ) to give cyclopropane derivatives. The reaction involves addition of carbene to C_2—C_3 bond of thiophen.

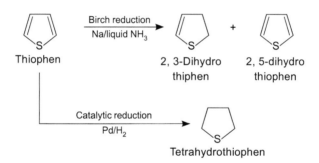

(*v*) **Lithiation:** Lithiation of 2, 4-dibromothiophen with *n*-BuLi gives 2-lithio derivative (the reactivity of 2-bromo is higher than that of 4-bromo), which on treatment with CO_2 followed by acid gives the corresponding carboxylic acid.

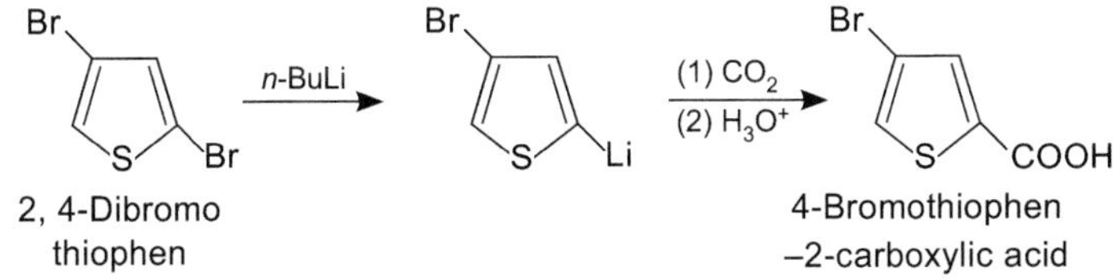

2, 4-Dibromo
thiophen

4-Bromothiophen
−2-carboxylic acid

(*vi*) **Reaction with diazonium sales:** Unlike furan and pyrrole (which undergo diazo coupling on reaction with diazonium salts), thiophen does not undergo coupling, but undergo arylation under alkaline conditions.

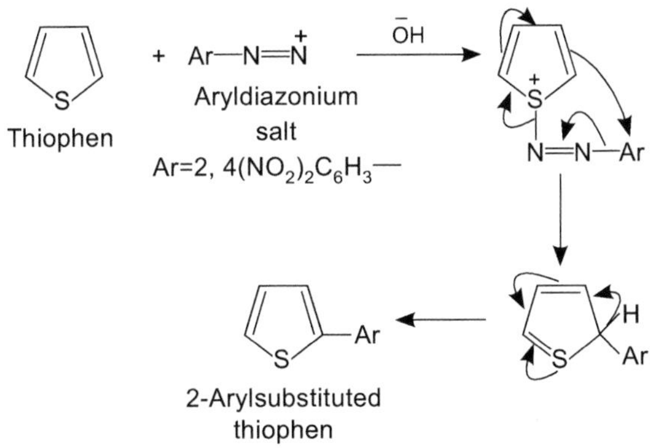

Thiophen

+ Ar—N≡N⁺
Aryldiazonium
salt
Ar=2, 4$(NO_2)_2C_6H_3$—

2-Arylsubstituted
thiophen

9.5   CONDENSED FIVE-MEMBERED HETEROCYCLES

These heterocycles contain a benzene ring fused to furan, pyrrole and thiophen at 2, 3-position giving rise to benzofuran, benzopyrrole (indole) and benzothiophen respectively.

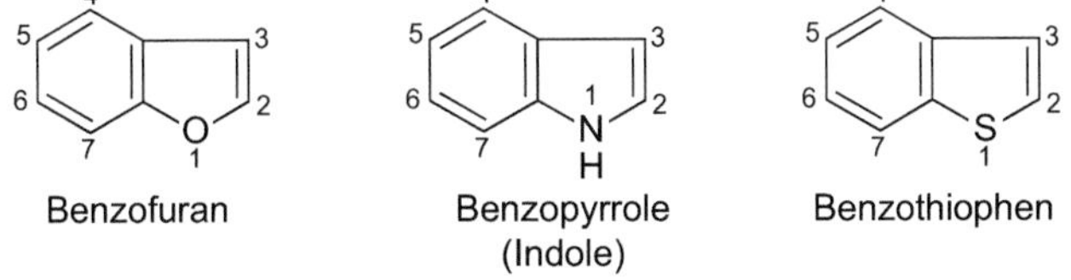

Benzofuran

Benzopyrrole
(Indole)

Benzothiophen

In the following section, only the chemistry of indole is discussed.

9.5.1 Indole

9.5.1.1 Synthesis

(i) *Fischer Indole Synthesis:* It involves cyclisation of phenyl hydrazones of an aldehyds or ketone in presence of acidic catalysts like H_2SO_4, $ZnCl_2$ etc.

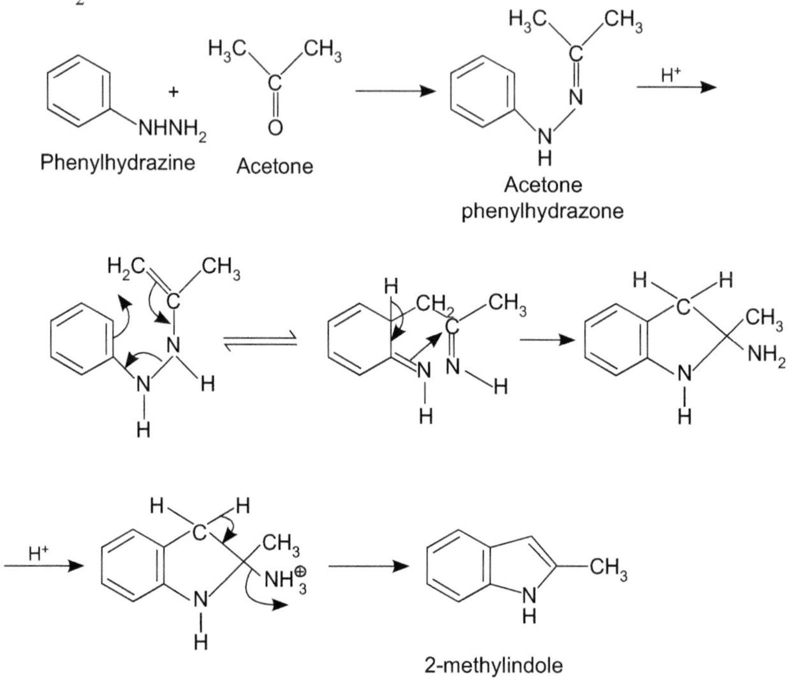

Phenylhydrazine   Acetone

Acetone
phenylhydrazone

2-methylindole

This is a general method for the synthesis of 2-substituted indoles. However, indole cannot be synthesized by this method. Indole can be prepared by cyclisation of phenylhydrazone of pyruvic acid followed by decarboxylation of the formed indole 2-carboxylic acid.

Phenylhydrazone
of pyruvic acid

Indole-2-
carboxylic acid

Indole

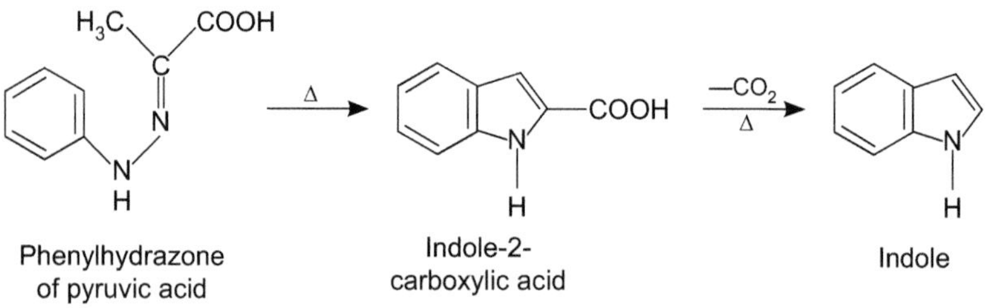

(ii) *Madelung Indole Synthesis:* It involves cyclodehydration of o-formyl-amidotoluene and o-acetamidotoluene with strong base to yield indole and 2-methyl indole respectively.

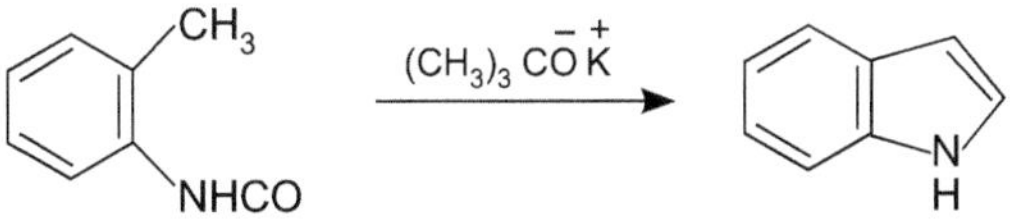

o-formylamido
toluene

Indole

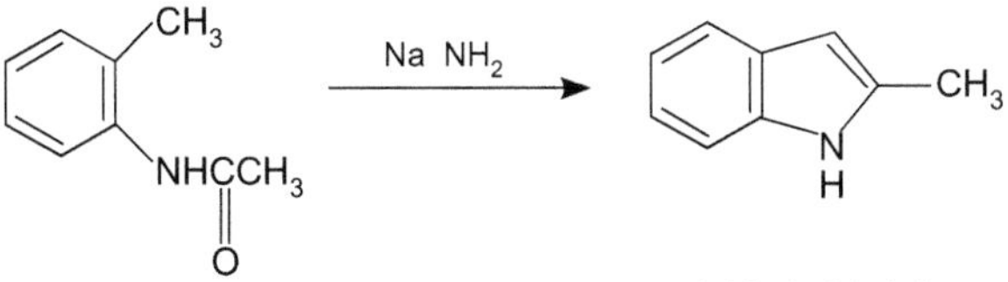

o-acetamido
toluene

2-Methyl-indole

(*iii*) **Bischler Indole Synthesis:** It involves heating phenacyl bromide with excess of aniline to give 2-phenylinidole.

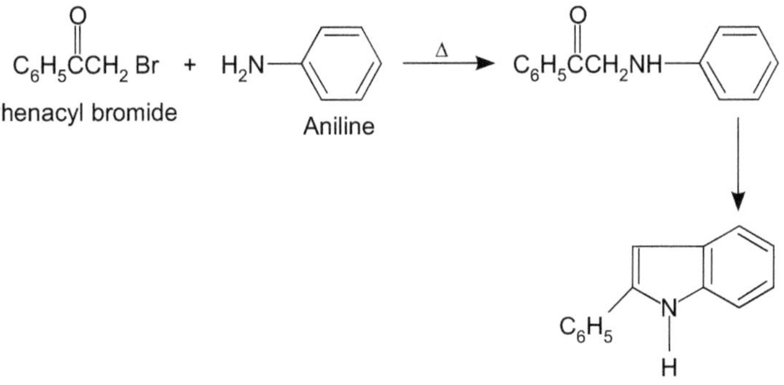

Phenacyl bromide

Aniline

2-phenylindole

(*iv*) **Reissert Indole Synthesis:** Condensation of o-nitrotoluene with diethyloxalate in presence of a basic catalyst followed by reduction of the formed ethyl o-nitro phenylpyruvic acid gives o-aminophenyl pyruvic acid. Finally, cyclodehydration of o-aminophenyl pyruvic acid gives indole-2-carboxylic acid, which on heating undergoes decarboxylation to give indole.

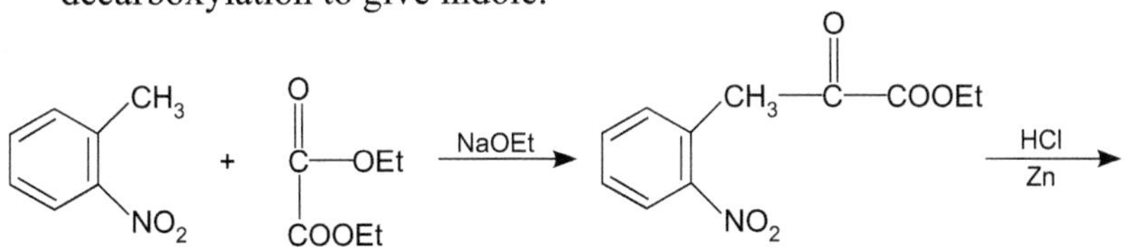

o-Nitrotoluene

Diethyl
oxalate

Ethyl o-nitrophenyl
pyruvic acid

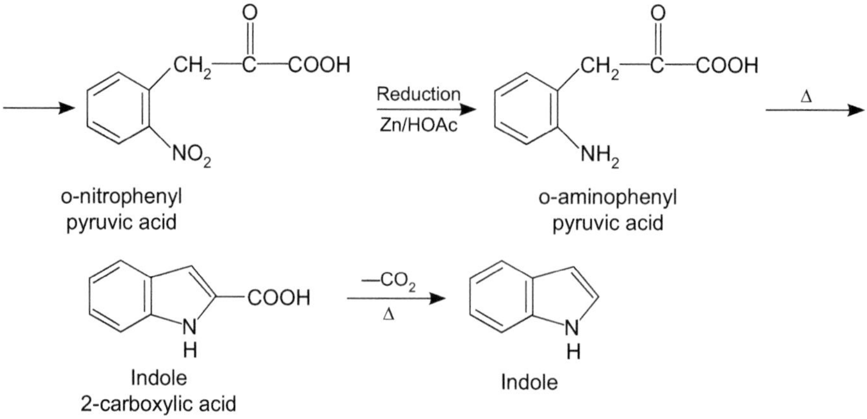

o-nitrophenyl
pyruvic acid

Reduction
Zn/HOAc

o-aminophenyl
pyruvic acid

Δ

Indole
2-carboxylic acid

−CO₂
Δ

Indole

9.5.1.2 Properties

Indole is a colourless crystalline solid, m.p 52 °C. It has 10 π-electrons over the cyclic frame work. The resonance energy of indole has been found to be 47.49 Kcal/mole. It undergoes following reactions.

(*i*) **Electrophilic substitution:** Contrary to electrophilic substitution in pyrrole, which lakes place at position 2, in case of indole the substitution takes place at position 3. This is due to comparatives stabilities of the σ complex (transition states) resulting due to electrophile attack at position 2 and 3.

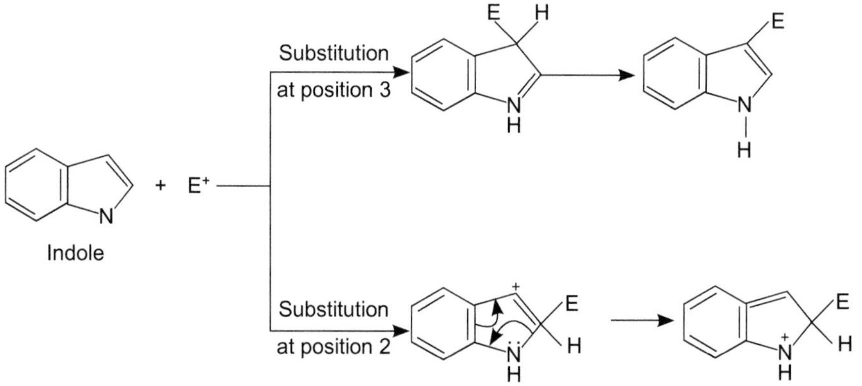

Some of the common electrophilic substitution reactions of indole are given below:

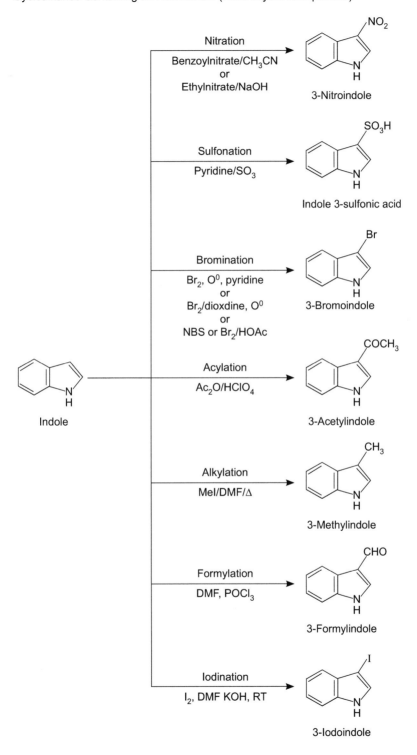

(*ii*) **Nucleophilic Substitution:** Due to excessive π-electron character nucleophilic substitution of indole is difficult. Dichlorocarbene (generated in situ) adds on to 2, 3-double bond of indole to give 3-chloro quinoline.

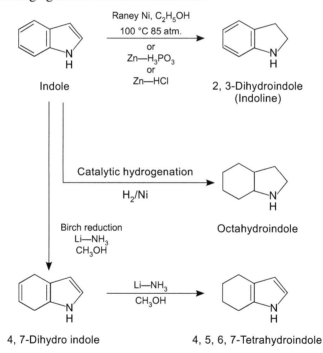

Indole, R = H
3-Methylindole, R = CH₃

3-Chloroquinoline, R = H
3-Chloro-4-methylquinioline
R = CH₃

(*iii*) **Reduction:** Indole on reduction gives products depending on the reducing agent and reaction conditions.

Raney Ni, C₂H₅OH
100 °C 85 atm.
or
Zn—H₃PO₃
or
Zn—HCl

Indole

2, 3-Dihydroindole
(Indoline)

Catalytic hydrogenation
H₂/Ni

Octahydroindole

Birch reduction
Li—NH₃
CH₃OH

Li—NH₃
CH₃OH

4, 7-Dihydro indole

4, 5, 6, 7-Tetrahydroindole

(*iv*) **Acidic Charactor:** As in the case of pyrrole, the hydrogen of N—H is acidic and can be replaced by metallic Na, K or Li. Also reaction with grignarel reagent gives indolyl magnesium bromide. The lithium derivation on treatment with CO_2 followed by acidification gives indole-1-carboxylic acid.

Indole — KOH 125–130°C → Indolyl potassium

Indole — RMgBr → Indolyl magnesium bromide

Indole — n-C$_4$H$_9$Li either, Δ → Indolyl lithium — (1) CO$_2$ (2) H$_3$O$^+$ → Indole-1-carboxylic acid

(v) **Basic Character:** Indole like pyrrole is a weak base. Thus, in acidic medium, indole gets protonated at position 3. This is followed by condensation with another molecule of indole to give the dimer.

Indole — H$^+$ → — Indole → Dimer

(vi) **Oxidation:** Indole on oxidation by air give indoxyl (indolin-3-one), which on further oxidation gives indigo.

Indole — O$_2$ air → — → Indoxyl — O$_2$/air → Indigo

Peroxyacids or ozone oxidise indole to gives 2-formamido benzaldehyde

Indole — C$_6$H$_5$CO$_3$H → 2-Formamidobenzaldehyde

9.6 SIX MEMBERED HETEROCYCLIC COMPOUNDS

Typical six-membered heterocyclic compounds include α-pyran, γ-pyran and pyridine.

α-Pyran   γ-pyran   Pyridine

We will discuss only the chemistry of pyridine.

9.6.1 Pyridine

9.6.1.1 Introduction

Pyridine is an important solvent and raw material for a number of industrially important substances.

The pyridine molecule is flat with bond angles of 120° and the carbon-carbon and carbon-nitrogen bond lengths are intermediate between those for single and double bonds

C—C bond length in pyridine 1.39 Å	C—C bond length 1.54 Å
	C==C bond length 1.34 Å
C—N bond length in pyridine 1.37 Å	C—N bond length 1.47 Å
	C==N bond length 1.28 Å

The stabilization energy of pyridine is 21 Kcal/mole. Like benzene, it resists addition and undergoes electrophilic substitution. These observations indicate that pyridine molecule has an aromatic sexet. All the five carbons and nitrogen atom are sp^2 hybridized. The unhybridized p-orbital at each carbon and nitrogen atom are perpendicular to the plane of the ring atoms and overlap side ways with each other to form π-electron cloud above and below the plane of the ring. The molecular orbital picture of pyridine is given below:

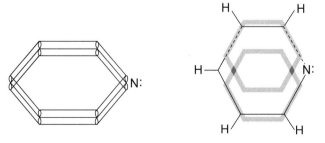

Pyridine is a stronger base than pyrrole. This is because the sp^2 hybrid orbital of nitrogen in pyridine contains a pair of electrons which become available for bonding with acids. The structure of pyridine is best described as a resonance hybrid of the following contributing forms.

As seen, in pyridine, the electronegative nitrogen produce a deficiency of electrons in the ring compared to pyrrole, where the ring carbon acquire increased electron density. If can, thus be said that the nitrogen atom in pyridine deactivates the ring.

9.6.1.2 Synthesis

(*i*) Pyridine is found to occur in coal tar along with some methylpyridines (picolines). It is commercially obtained from light oil fraction by extraction with dilute H_2SO_4, which dissolves pyridine and picolines. The acidic extract is basified with NaOH and the separated oil on fractional distillation gives pyridine, b.p. $115 - 116\ ^\circ C$.

(*ii*) **From   pentamethylene   diamine:**   Pentamethylene   diamine dihydrochloride on heating gives piperidine, which on catalytic dehydrogenation yields pyridine.

Pentamethylene   Piperidine   Pyridine
diamine dihydrochloride

(*iii*) **From acetylene and hydrogen cyanide:** A mixture of acetylene and hydrogen cyanide on passing through red hot tube gives pyridine.

(*iii*) **Hantzsch synthesis:** It involves condensation of a β-ketoester, an aldehyde and ammonia to gives dihydropyridine, which on oxidation gives the corresponding pyridine derivative.

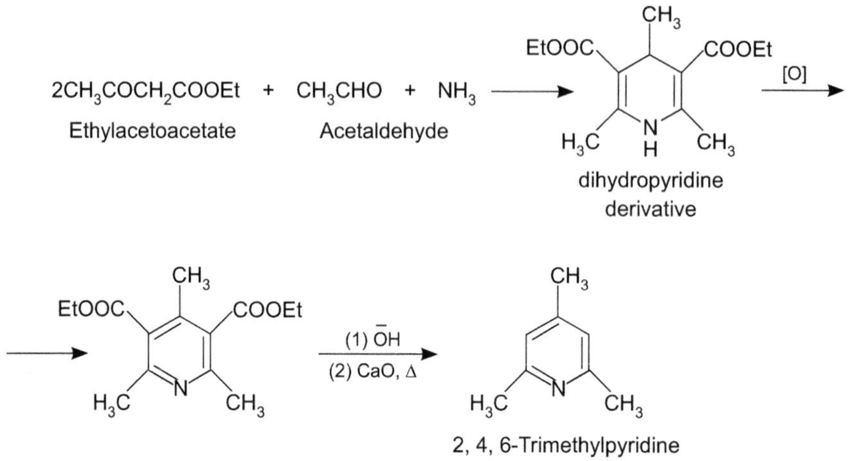

2CH₃COCH₂COOEt + CH₃CHO + NH₃ ⟶

Ethylacetoacetate   Acetaldehyde

dihydropyridine
derivative

[O] ⟶

$\xrightarrow{\text{(1) } \overline{O}H}$
$\xrightarrow{\text{(2) CaO, } \Delta}$

2, 4, 6-Trimethylpyridine

Using this method a wide variety of pyridine derivatives can be prepared.

The reaction is believed to take place in the following three steps.

(*i*) CH₃COCH₂COOEt + NH₃ $\xrightarrow{-H_2O}$ CH₃C=CHCOOEt

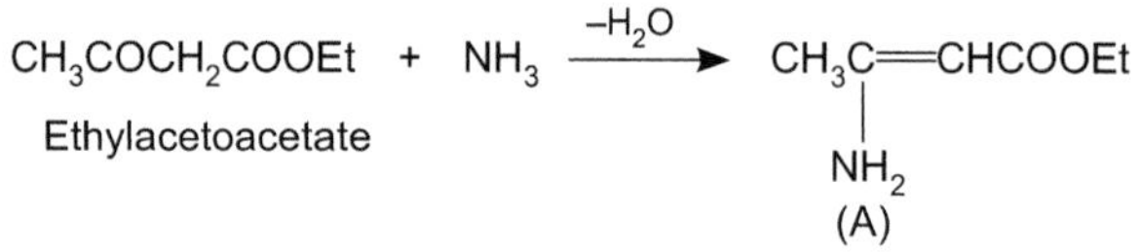

Ethylacetoacetate

NH₂
(A)

(*ii*) CH₃—C=O + H₂C(COCH₃)(COOEt) $\xrightarrow{NH_3}$ H₃C—C=C(COCH₃)(COOEt)

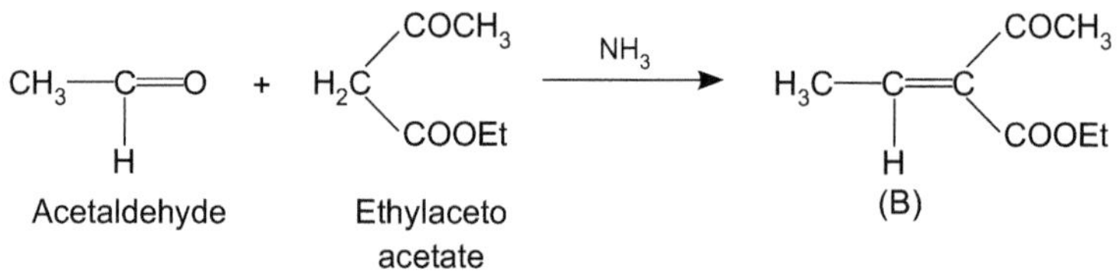

Acetaldehyde   Ethylaceto
acetate

(B)

(*iii*)

Michael type
addition ⟶

(A)   (B)

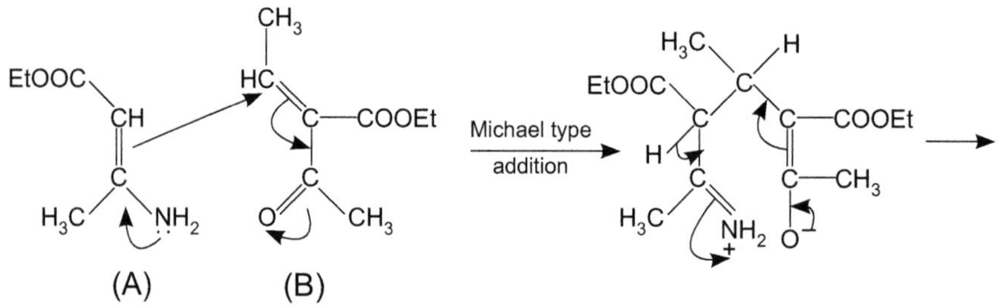

(iv) From pyrrole see page 121

9.6.1.3 Properties

Pyridine is a much stronger base than pyrrole but is much weaker than aliphatic amines. That pyridine is a much stronger base than pyrrole can be rationalised by considering the fact that pyridine nitrogen has an electron pair which is not required for stablization of the aromatic system, which can accept a proton to form the corresponding conjugate acid, the pyridinium ion. On the other hand, the electron pair on nitrogen atom in pyrrole is not available for sharing with acids. In fact, pyrrole can accept proton only at the expense of the aromatic character of the ring.

The reason for lesser basic strength of pyridine compared to aliphatic amines is well understood from the following reasoning. The pair of electrons which is responsible for base strength of pyridine is held in an sp^2 orbital (the N is sp^2 hybridized). On the other hand the pair of electron on N in aliphatic amines is in sp^3 orbital. It is well known that the base strength of a substance depends on the availability of this unshared pair for sharing with acids. Also an electron in an s orbital is held more tightly due to its proximety to the nucleus compared to an electron in a p-orbital. As the proportion of s-character is more in sp^2 orbital then in sp^3 orbital, the pair of electron in nitrogen of pyridine (sp^2 hybridization) is some what less available for sharing with acids than in the pair of electrons in an alephatic amines (sp^3 hybridization).

On the basis of the above, the basicity of pyridine, pyrrole and aliphatic amines of is of the order.

aliphatic amins > pyridine > pyrrole

Pyridine undergoes following reactions.

(*i*) *Electrophilic substitution:* Pyridine ring is deactivated for electrophilic substitution due to the withdrawal of electrons from the ring towards nitrogen atom (see resonating structures of pyridine, page 135).

There is another factor which is responsible for deactivity of pyridine ring. It is that under acidic conditions in most of the electrophilic substitutions (like nitration, sulphonation etc.) pyridinium ion is formed, which repels the positively charged electrophiles. So vigorous reaction conditions are necessary for electrophilic substitutions.

In pyridine, the N atom deactivates positions 2 and 4 more than the position 3. So electrophilic substitution takes place preterably at position 3. This is well understood if we consider the structures arising due to attack at position at 4, 3 and 2.

As seen, attack at position 4 and 2 involves a structure in each bearing positive charge on nitrogen (structures marked A and B respect), whereas there is none of the intermediate structures formed from attack at position 3 bear a positive charge on nitrogen; so the intermediates formed by electrophilic attack at position 3 is more stable. Therefore electrophilic substitution is favoured at position 3.

Typical electrophilic substitution reactions in pyridine are nitration, sulfonation and bromination. All these reactions require drastic conditions due to deactivation of the ring.

Nitration
KNO₃, H₂SO₄
370 °C → 3-Nitropyridine (major) + 2-Nitropyridine (minor)

Sulfonation
SO₃—H₂SO₄
HgSO₄, 200 °C → Pyridine-3-sulphonic acid

Bromination
Br/350 °C → 3-Bromopyridine + 3,5-dibromo-pyridine

(*ii*) *Nucleophilic Substitution:* In pyridine, nucleophilic substitution takes place at position 2 and 4. This is well understood by consideration of the intermediates arising due to attack of nucleophile on position 2, 3 and 4. Substitution does not take place at position 3.

As seen, one of the structures obtained due to attack at position 2 and 4 there is negative charge on N (structures marked C and D respectively). There is no such structure arising due to attack at position 3. So the nucleophilic attack occurs at position 2 and 4. Some examples of nucleophilic substitution of pyridine are given below.

A typical example of nucleophilic substion is the reaction of pyridine with NaNH₂ give 2-amino pyridine (**Tschitschibabin reaction**).

Pyridine + NaNH₂ → [intermediate] → 2-aminopyridine + NaH

—H₂

2-aminopyridine ← H₂O ← NHNa derivative

Some other nucleophilic substitution reactions and given below:

$C_6H_5^{\delta-}Li^{\delta+}$ → 2-phenylpyridine + LiH

KOH 320 °C → 2-pyridinol ⇌ 2-pyridone

$n\text{-}C_4H_9\,Li$ → 2-Butylpyridine + LiH

As seen, the presence of N in pyridine activates the ring for nucleophilic substitution. This is also examplified by the facile displacement of halogen in 2- and 4-positions.

2-Bromopyridine $\xrightarrow[200\,°C]{NH_3}$ 2-aminopyridine

4-Chloropyridine $\xrightarrow[200\,°C]{NH_3}$ 4-aminopyridine

The 3-halopyridine, as expected are inert to direct substitution reactions. However 3-halopyridines are attacked by a strongly basic nucleophiles like

KNH_2/NH_3. In this case, the products are formed due to the intermediate formation of benzyne like intermediates.

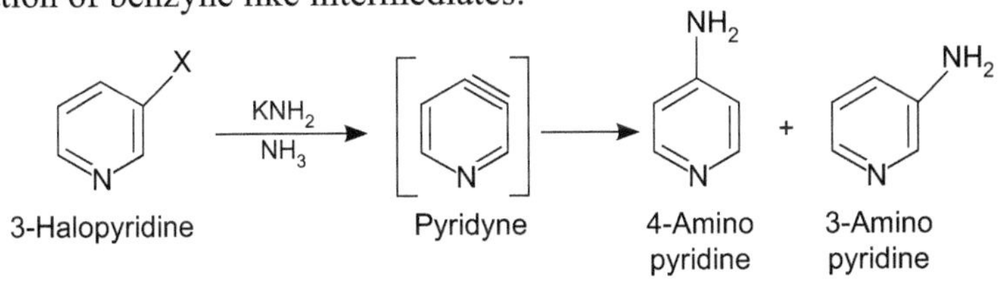

3-Halopyridine   Pyridyne   4-Amino   3-Amino
                                                     pyridine   pyridine

(*iii*) *Aldol-type condensation:* The methyl group in position 2 and 4 are relatively acidic due to electronegativity of N in pyridine and so undergo aldol-type reactions with aldehydes.

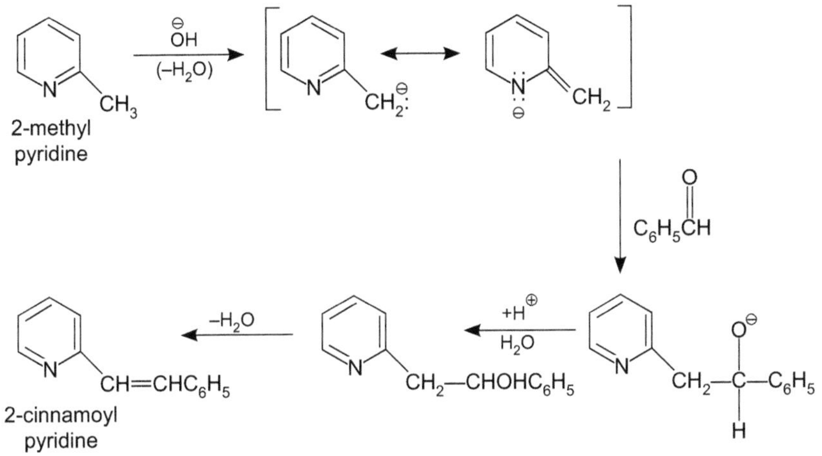

2-methyl
pyridine

$C_6H_5\overset{O}{\overset{\|}{C}}H$

2-cinnamoyl
pyridine

The reactivity of α– γ-picolines can be increased by conversion to the corresponding pyridinium salt.

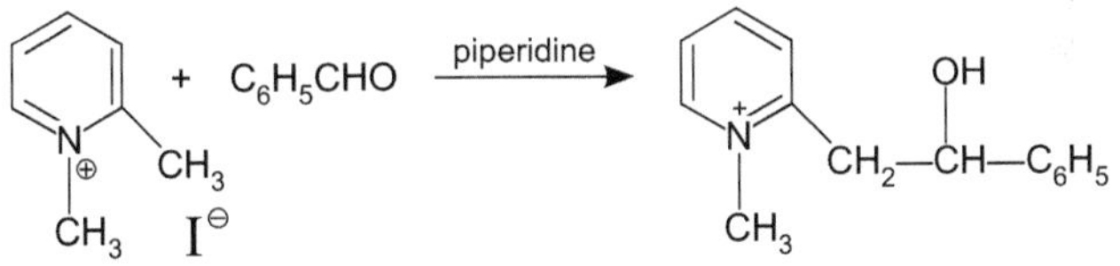

(*iv*) *Reduction:* Catalytic hydrogenation of pyridine gives piperidine.

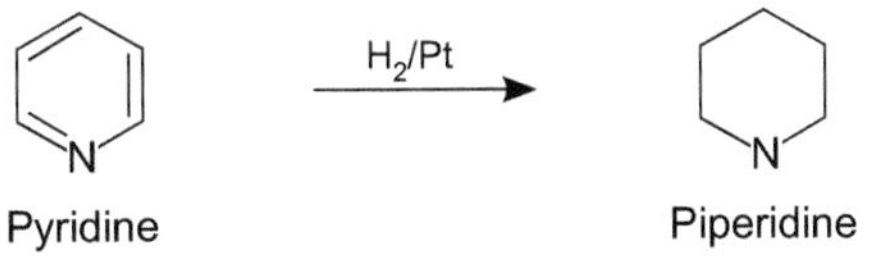

Pyridine   Piperidine

In piperidine, the nitrogen atom is sp^3 hybridized and is therefore a much stronger base than pyridine. It is used as a basic catalyst in a number of organic reactions. The piperidine ring is opened either by oxidation with H_2O_2 or by Holmann exhaustive methylation.

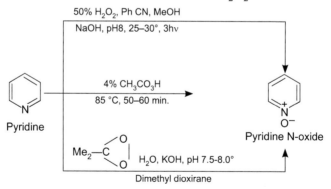

(v) **Oxidation:** Oxidation of pyridine with per acids give pyridine N-oxide.

Pyridine N-oxide is a very interesting compound. A discussion on pyridine N-oxide forms the subjection matter of a subsequent section.

9.6.2 Pyridine N-Oxide

As already stated, pyridine N-oxide is obtained by oxidation of pyridine with per acids. Best results are obtained by oxidation with dimethyl dioxirane, which is obtained in situ by the reaction of oxone ($2K_2SO_3$, $KHSO_4$, K_2SO_4) with acetone. Oxidation can also be effected with H_2O_2.

The presence of oxygen on nitrogen of the pyridine ring reverses the direction of electrophilic substitutions in the pyridine ring. We know that normally attack on pyridine ring occurs at β-position (position-3). However, in pyridine N-oxide, the electrophilic attack occurs at α(2 position) or γ (4 position) carbon. After introduction of the electrophile, the pyridine N-oxide is convested into pyridine by mild reduction like treatment with salts of iron or titanium or Zn/HOAC. In fact this is a convenient method for the synthesis of 2- or 4- substituted pyridines.

In pyridine N-oxide there are two powerful, but opposed polarisation effects. Pyridine N-oxide is represented by the following structures which contribute to the resonance hybrid.

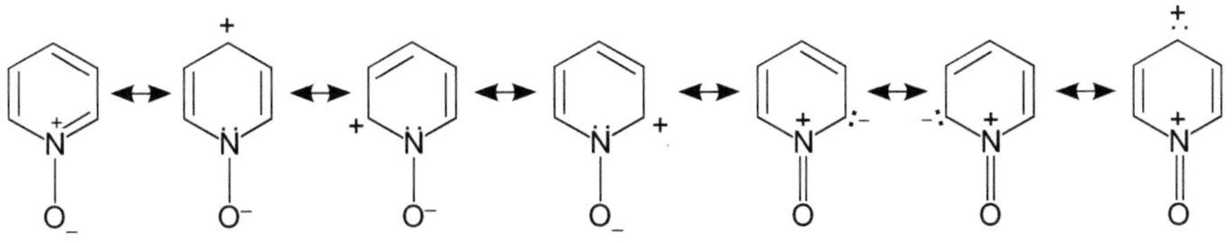

As seen in the above structures, in some forms there is decrease in electron density at various ring positions while in other forms there is an increase in electron density. Thus pyridine N-oxide undergo electrophilic and nucleophilic substitutions. Some of the reactions of pyridine N-oxide are given below:

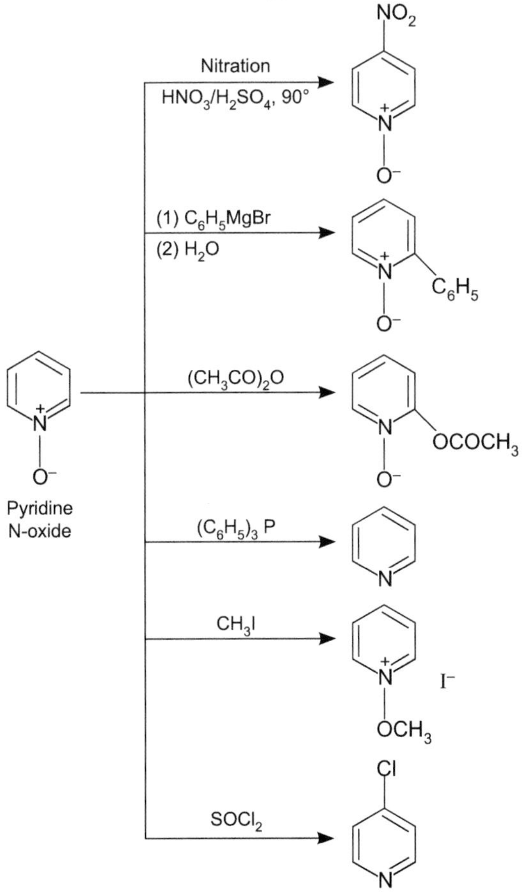

On the basis of the above it can be said that pyridine is much stronger base than pyrrole compared to piperidine, pyridine is less basic, pyrrole is least basic compared to pyridine and piperidine.

Piperidine > Pyridine > Pyrrole

9.6.3 Condensed Six-membered Heterocycles

Typical examples of condensed six-membered hetrocycles are quinoline and isoquinoline.

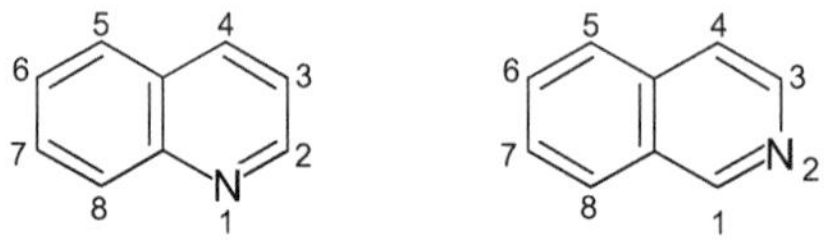

Quinoline   Isoquinoline

9.6.3.1 Quinoline

Quinoline is one of the degradation product of the akaloid quinine. It is obtained from coal tar.

9.6.3.2 Synthesis

(*i*) *Skraup synthesis:* A widely used procedure for the synthesis of quinoline and its derivatives, consists of heating a primary aromatic amine having one free ortho position with an α, β-unsaturated carbonyl compound (or its precursor, mostly glycerol) in presence of a condensing agent (e.g. conc. H_2SO_4) and an oxidant (nitribenzone). Quinoline can be obtained by heating a mixture of aniline, nitrobenzene, glycerol and conc. H_2SO_4 and $FeSO_4$.

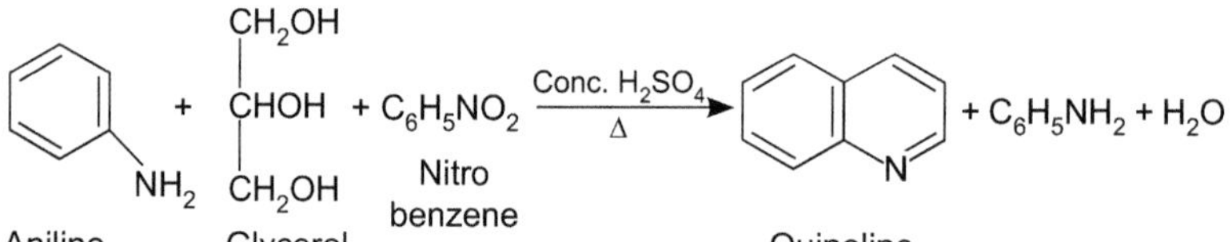

Aniline   Glycerol   benzene   Quinoline

The function of $FeSO_4$ in the above reaction is to make it less violent.

The skraup synthesis is believed to take place in four steps.

(*a*) Dehydration of glycerol into acrolein.

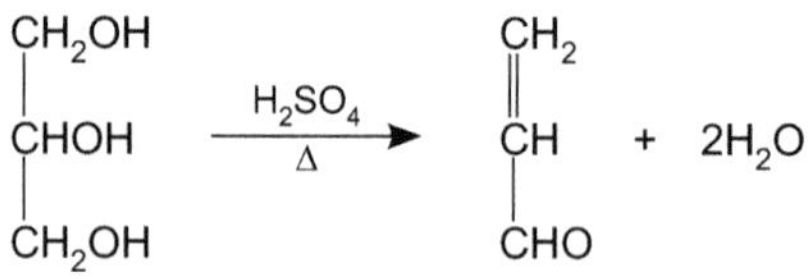

Glycerol   Acrolein

(b) Michael addition of aniline to acrolein to yield β-(phenylamino) propionaldehyde.

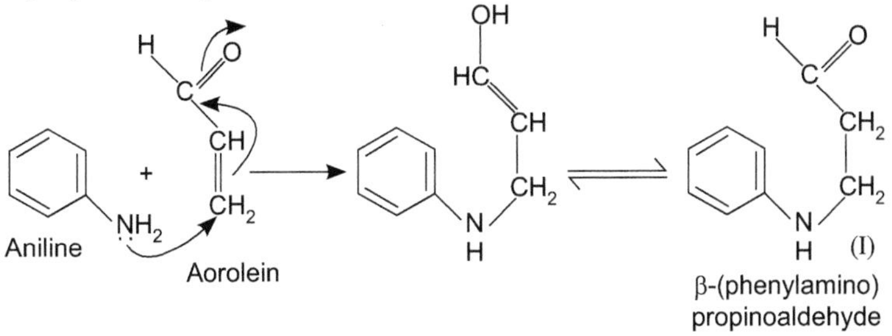

(c) Intramolecular electrosubstitution upon the aromatic ring by protonated aldehyde followed by dehydration to give 1, 2-dihydroquinoline.

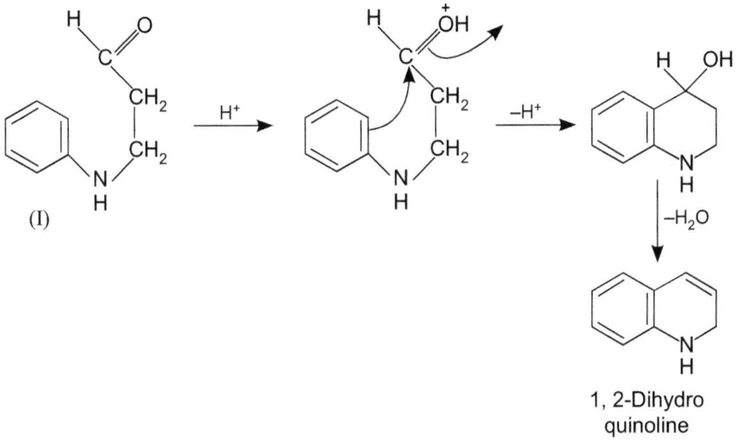

1, 2-Dihydro quinoline

(d) Dehydrogenation of 1, 2-dihydroquinoline to quinoline.

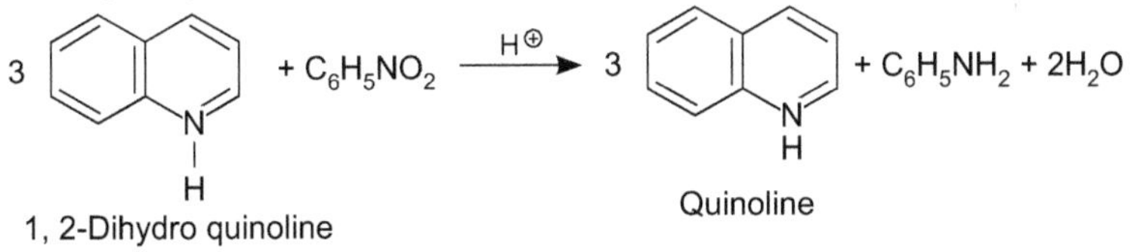

(ii) **Doebner-Miller Synthesis:** It is a modification of skarup synthesis and is useful for the synthesis of quinoline homologues. The method consist in using an aldehyde or a ketone in place of glycerol. The reaction starts with an initial formation of α, β-unsaturated carbonyl compound, subsequent steps are same as in skarup synthesis. By using acetaldehyde, 2-methylquinoline is obtained.

$$C_6H_5NH_2 \ + \ O{=}CHCH_3 \longrightarrow C_6H_5N{=}CHCH_3$$

Aniline   Acetaldehyde   Schiffs base (A)

$$CH_3CHO \xrightarrow[\text{aldol condensation}]{H^+} \underset{\text{aldol}}{H_3C{-}\overset{\overset{\displaystyle OH}{|}}{CH}{-}CH_2CHO} \longrightarrow \underset{\text{2-Butenal}}{CH_3CH{=}CHCHO}$$

Acetaldehyde

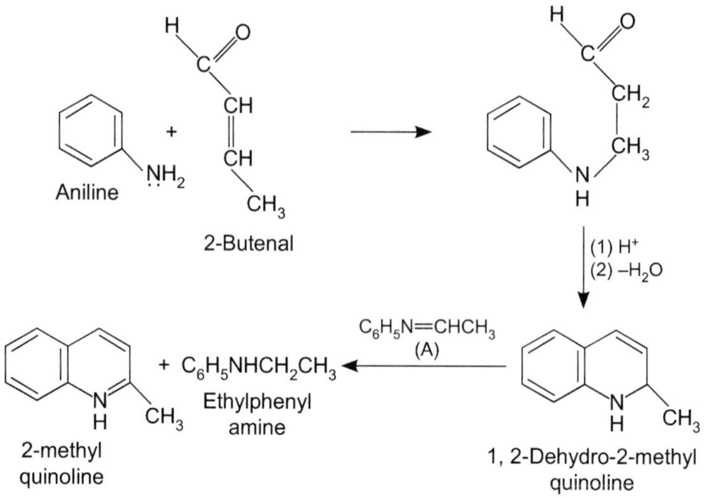

Aniline + 2-Butenal

2-methyl quinoline + C$_6$H$_5$NHCH$_2$CH$_3$ (Ethylphenyl amine)

C$_6$H$_5$N=CHCH$_3$ (A)

1, 2-Dehydro-2-methyl quinoline

(1) H$^+$
(2) –H$_2$O

(*iii*) *From Indole:* Treatment of indole with methyllithium in methylene chloride solution gives quinoline. In this reaction methyllithium reacts with methylene chloride to generate chloromethylene (in situ) which adds on to indole followed by ring expansion to give quinoline.

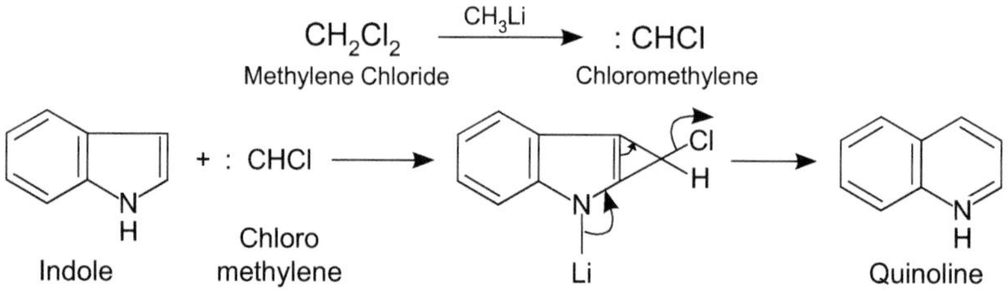

$$CH_2Cl_2 \xrightarrow{CH_3Li} :CHCl$$

Methylene Chloride   Chloromethylene

Indole + : CHCl (Chloro methylene) → (Li) → Quinoline

(*iv*) *Friedlander synthesis:* It involves condensation of an o-amino aromatic aldehydes or ketone with a carbonyl compound. The reaction proceeds via the formation of a schiff base followed by an internal aldol-type condensation between an aryl carbonyl and the activated methylene group.

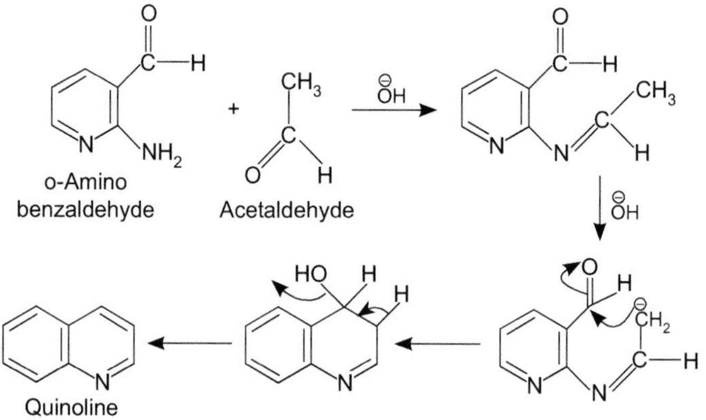

o-Amino benzaldehyde + Acetaldehyde $\xrightarrow{\overset{\ominus}{O}H}$

$\downarrow \overset{\ominus}{O}H$

Quinoline

The Friedlander synthesis is unsuitable for the synthesis of quinolines having substituents in the benzene ring because the required o-amino aromatic aldehydes or ketones are difficult to prepare. This problem has been overcome by using isatins (**Pfitzinger synthesis**). The reaction involves ring cleavage of isatin by base to yield isatic acid derivative, which undergoes condensation with carbonyl compounds to give quinoline derivatives.

2, 6-Dimethyl quinoline

(*v*) *Conrad-Limpach and Knorr quinoline synthesis:* The reaction of aniline with ethylacetoacetate may produce two products, viz, 2-quinoline or 4-quinoline depending on the conditions of the reaction.

2-Methyl-4-quinoline

4-Methyl-2-quinoline

The pathway (*a*) is known as **conrad-limpach synthesis** and pathway (*b*) Constitutes the **knorr quinoline synthesis**. The formation of these two products can be explained by the fact that the nucleophilic attack by aniline may occur at the carbonyl group of ketone or ester grouping resulting in the formation of enamine or anilide respectively. Cyclisation of enamine at 250°C produces 4-quinolone and heating anilide with conc. H_2SO_4 gives 2-quinoline.

(a) Formation of 4-quinoline

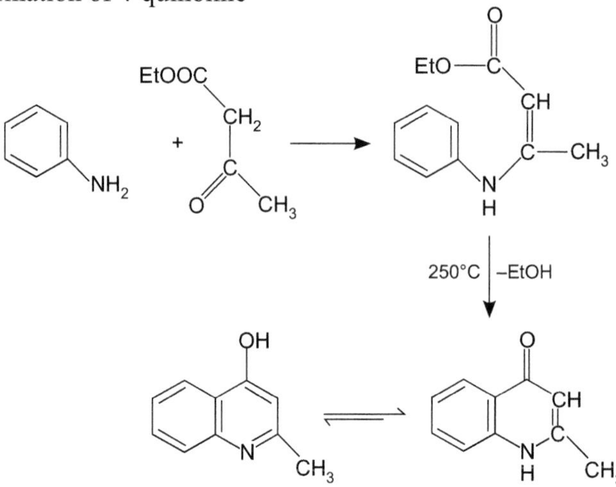

2-methyl-4-quinolone

(b) Formation of 2-quinolone

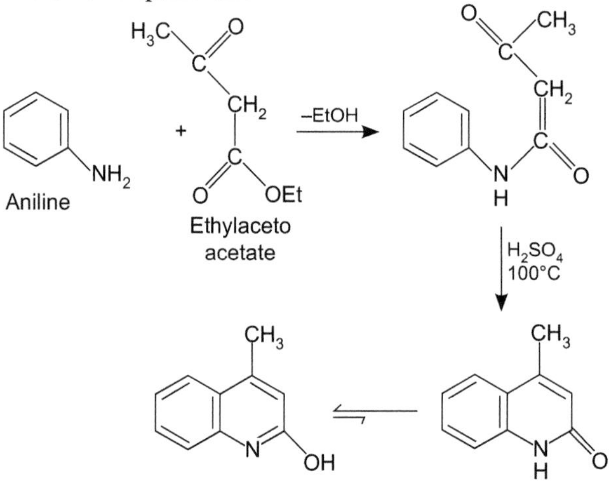

Aniline

Ethylaceto acetate

4-methyl-2-quinolone

Substituted quinolines can also be obtained by using 1, 3-diketones in place of β-keto esters.

9.6.3.3 Properties

Quinoline, b.p. 238° is basic in nature. The principal resonance form of quinoline are:

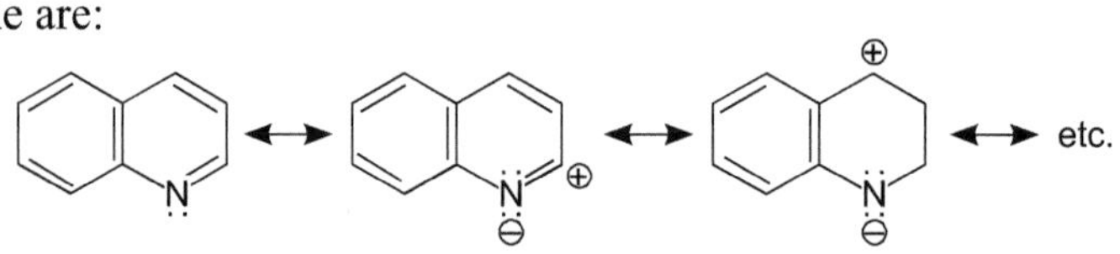

Some of the important reactions of quinoline are given below:

(*i*) **Basic character:** As in the case of pyridine, quinoline has a lone pair of electrons on N, which does not contribute to the aromatic sexet. Being a tertrary amine, quinoline forms salts with in organic acids. With CH_3I it forms N-methyl quinolium iodide.

Quinoline   Quinolium chloride

N-methylquinolium
iodide

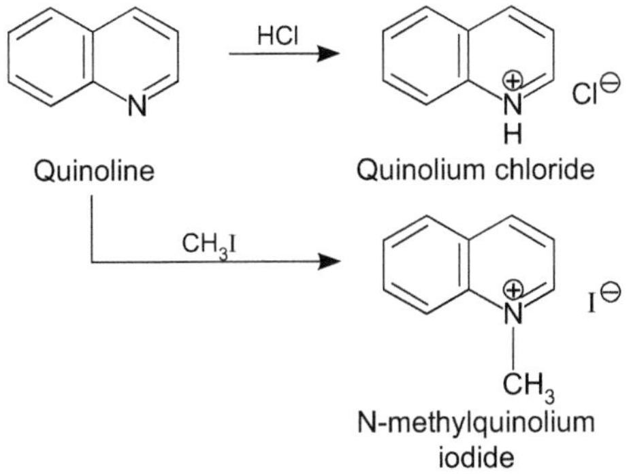

(*ii*) **Electrophilic substitution:** Quinoline (like pyridine) undergoes electrophilic substitution. The nitrogen containing ring in quinoline resembles pyridine and the carbocyclic ring has greater resemblance with that of benzene. In fact, electrophilic substitutions in quinoline can be effected using less vigorous conditions than that required for pyridine. Some of the typical electrophilic substitution reactions of quinoline are given below.

5-Nitroquinoline   8-Nitroquinoline

Quinoline

5-Bromoquinoline   8-Bromoquinoline

Quinoline
8-sulphonic
acid

8-Hydroxyquinoline

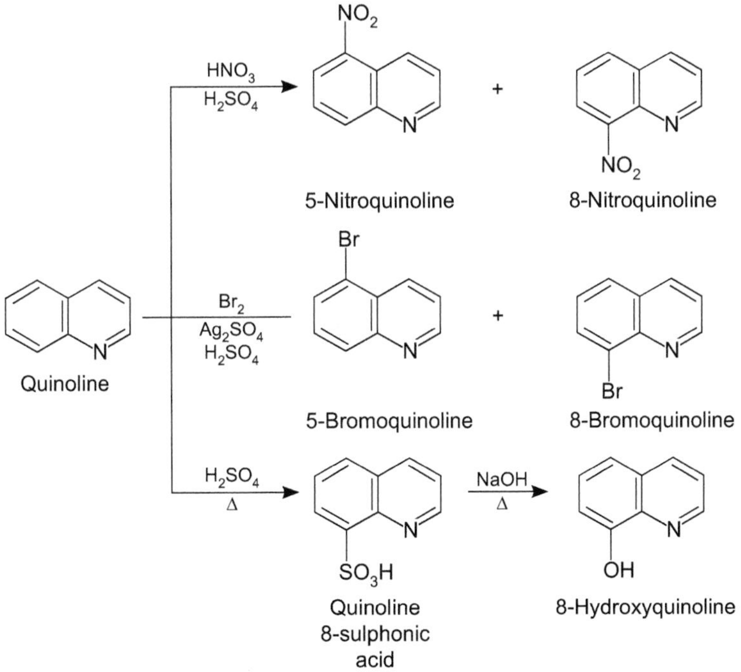

As seen quinoline on nitration gives a mixture of 5 and 8-nitroquinoline. However, 4-nitroquinoline is obtained from quinoline by reacting with peracid followed by nitration of the formed quinoline N-oxide to give the corresponding 4-nitro product. Final reduction with Sn/HCl gives 4-nitroquinoline.

Sulphonation of quinoline with H_2SO_4 at 300° gives quinoline 6-sulphone acid which can also be obtained from quinoline 8-sulfonic acid by heating at 300°.

(*iii*) *Nucleophilic substitution:* Like pyridine, quinoline also undergoes nucleophilic substitution. Thus, quinoline reacts with sodium amide to give 2-aminoquinoline, which on subsequent treatment with HNO_2 gives crbostyril. Similarly, treatment of quinoline with *n*-butyllithium gives 2-butyl derivative.

If 2-position of quinoline is occupied, nucleophilic attack takes place at position 4. An eample is given below.

(iv) **Reduction:** Reduction of quinoline give product depending on the reducing agent.

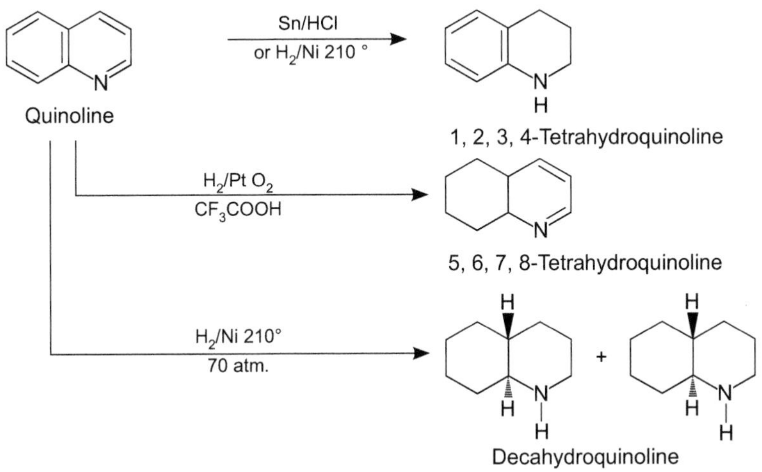

Quinoline

Sn/HCl or H₂/Ni 210 °

1, 2, 3, 4-Tetrahydroquinoline

H₂/Pt O₂
CF₃COOH

5, 6, 7, 8-Tetrahydroquinoline

H₂/Ni 210°
70 atm.

Decahydroquinoline

(v) **Oxidation:** The product obtained by oxidation of quinoline depends on the oxidizing agent and the reaction conditions.

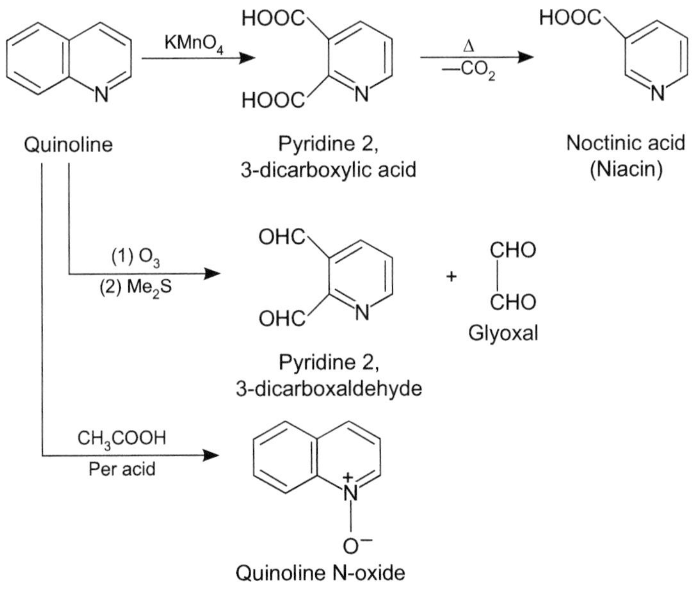

Quinoline

KMnO₄

Pyridine 2, 3-dicarboxylic acid

Δ
−CO₂

Noctinic acid (Niacin)

(1) O₃
(2) Me₂S

Pyridine 2, 3-dicarboxaldehyde

Glyoxal

CH₃COOH
Per acid

Quinoline N-oxide

(vi) **Reaction of 4-methyl quinoline with benzaldehyde:** Alkyl groups in ortho or para position to the N in quinoline are acidic and react with aldehydes to give the corresponding styryl quinoline.

4-methyl quinoline   4-styrylquinoline

9.6.4 Isoquinoline

Isoquinoline is found to occur in coal tar and bone oil. It is one of the few compounds in which numbering of the ring does not start with the hetero atom

Isoquinoline is a decomposition product of many alkaloids. It is obtained commercially from coal tar and bone oil.

9.6.4.1 Synthesis

(*i*) *Bischler-Napieralski Synthesis:* It involves the reaction of β-phenyl ethylamine (obtained by the reaction of benzaldehyde with nitromethane in presence of base, followed by reduction of the formed ω-nitro styrene) on acylation and subsequent heating with acidic reagent undergoes cyclodehydration to give 3, 4-dihydroisoquinoline. Final aromatisation gives the corresponding isoquinoline.

Benzaldehyde   ω-nitrostyrene   β-phenylethylamine

3, 4-Dihydro   Isoquinoline
isoquinoline

(*ii*) **Pictet-Spenler Synthesis:** The method involves condensation of β-phenyl ethylamine (obtained as given above) with an aldehyde followed by cyclisation of the formed schiffs base in presence of acid yield tetrahydroisoquinoline. Final dehydrogenation give the corresponding isoquinoline.

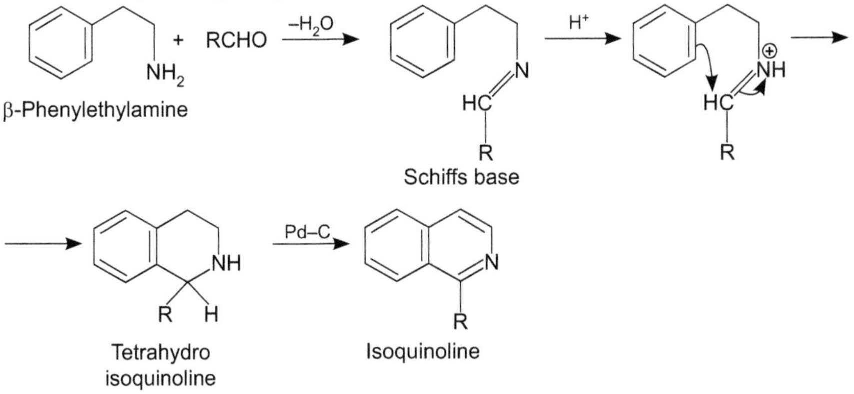

β-Phenylethylamine

Schiffs base

Tetrahydro isoquinoline

Isoquinoline

(*iii*) **Pomeranz-Fritsch Synthesis:** Treatment of an aromatic aldehyde with an aminoacetal followed by cyclization of the formed schiffs base in presence of acid yields isoquinoline.

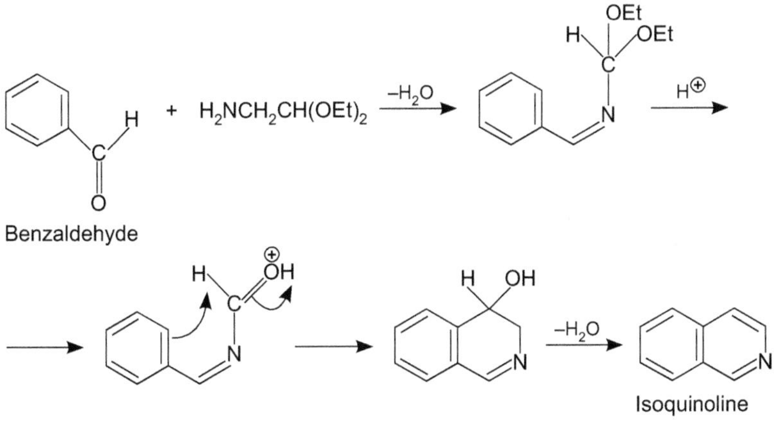

Benzaldehyde

Isoquinoline

9.6.4.2 Properties

Isoquinoline, b.p. 243° is basic in nature. The principal resonance forms contributing to hybrid are given below:

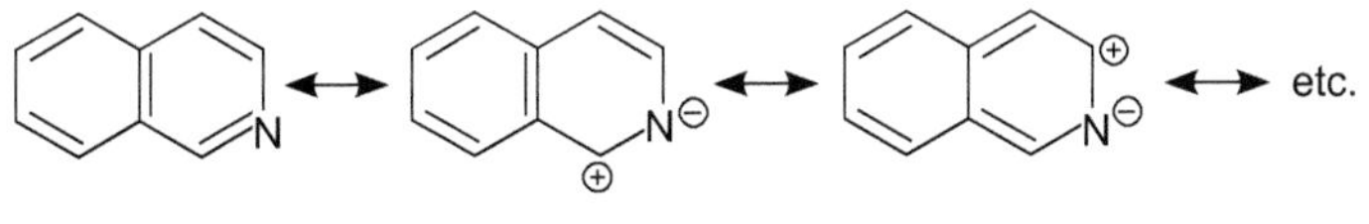

etc.

(A)

Form A contributes to a great extent, as in the other forms the benzenoid character of isoquinoline is destroyed.

(i) **Flectrophilic Substitution:** Like quinoline, isoquinoline also undergoes electrophilic substitution prefererbly at C-5 and C-8 positions. Some of the electrophilic substions of isoquinoline are shown below:

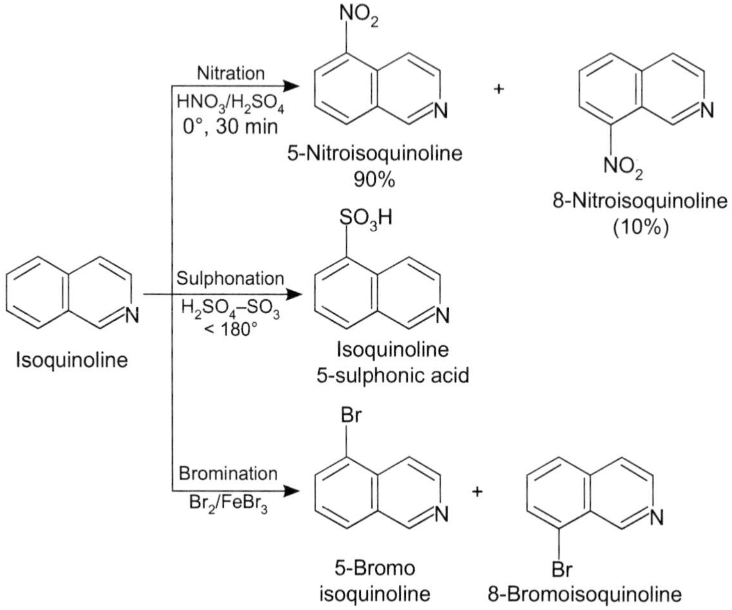

5-Nitroisoquinoline
90%

8-Nitroisoquinoline
(10%)

Isoquinoline
5-sulphonic acid

5-Bromo
isoquinoline

8-Bromoisoquinoline

Isoquinoline

(ii) **Nucleophilic substituion:** Isoquinoline undergoes nucleophilic substition at position 1. Thus, the reaction of NaNH$_2$/Liq. NH$_3$ (**Tschitschibabin reaction**) gives 1-aminoisoquinoline, which can be converted into isocarbostyril on treatment with HNO$_2$.

Similarily, grignard reagent reacts with isoquinoline to give corresponding 1-alkyl substituted isoquinoline.

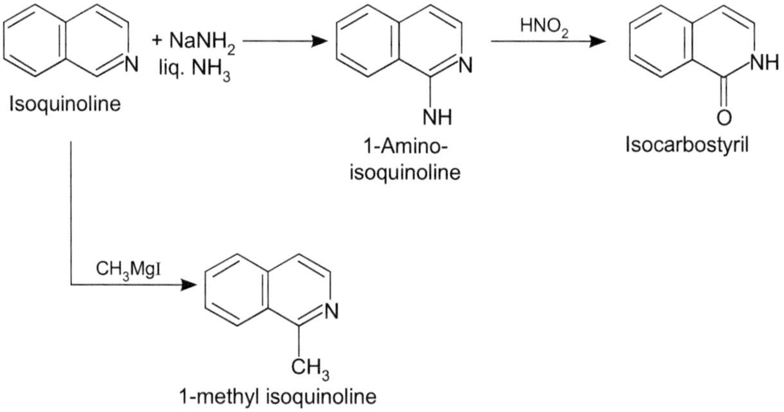

Isoquinoline

1-Amino-
isoquinoline

Isocarbostyril

1-methyl isoquinoline

(*iii*) **Oxidation:** Isoquinoline can be oxidised with $KMnO_4$ (alkaline or neutral) and with per acids.

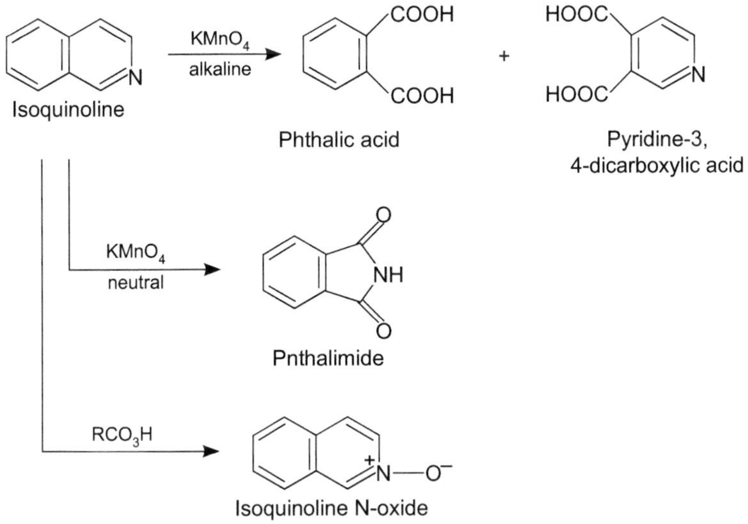

Isoquinoline → Phthalic acid + Pyridine-3, 4-dicarboxylic acid

Pnthalimide

Isoquinoline N-oxide

(*iv*) **Reduction:** Isoquinoline on reduction give products depending on the reducing agent and the reaction conditions.

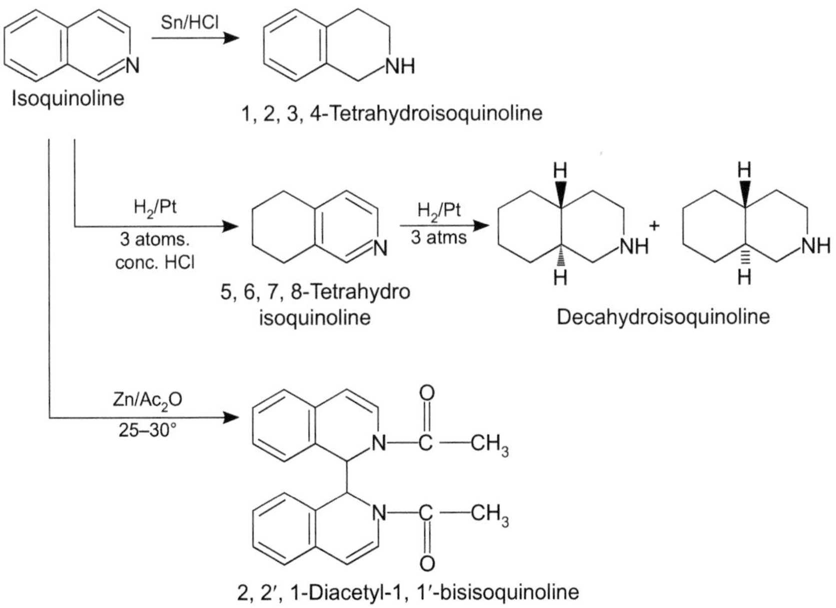

Isoquinoline → 1, 2, 3, 4-Tetrahydroisoquinoline

5, 6, 7, 8-Tetrahydro isoquinoline

Decahydroisoquinoline

2, 2', 1-Diacetyl-1, 1'-bisisoquinoline

(v) **Reaction of 1-methylisoquinoline with aromatic aldehydes:** Alkyl groups situated at orthoposition to nitrogen are acidic and react with aromatic aldehydes.

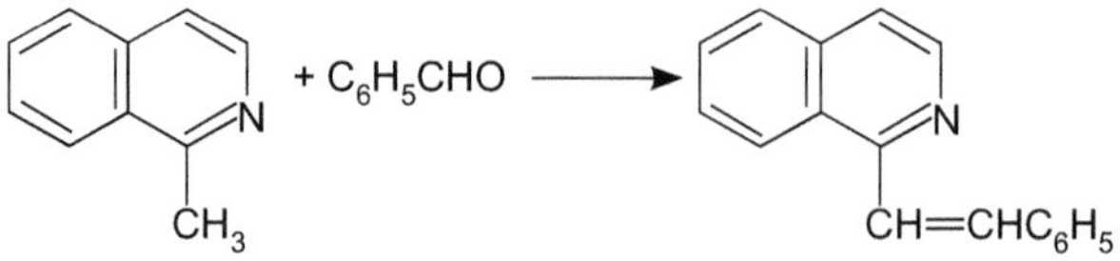

1-methyl
isoquinoline

1-Styrylisoquinoline

□□□

10

Non-Benzenoid Aromatics

10.1 INTRODUCTION

We know that Huckel rule (developed in 1931 by a German physicist Erich Huckel on the basis of mathematical calculations) is concerned with compounds containing one planar ring in which each atom has *a p* orbital as in benzene. On the basis of Huckel's calculations, it was shown that planar monocyclic rings containing (4n + 2) π-electrons, where n = 0,1,2,3..., and the rings containing 2,6,10,14... π-electrons have closed shells of delocalised electrons as in benzene, and should have substantial resonance energies.

It was, thus, concluded that planar monocyclic rings with 2,6,10,14... delocalised electrons should be *aromatic*.

10.2 AROMATIC, ANTIAROMATIC AND NON-AROMATIC COMPOUNDS

Aromatic compounds are those compounds in which the π-electrons are delocalised over the entire ring and that these compounds are stabilised by its π-electron delocalisation. The best way to find whether the π-electrons of a cyclic system are delocalised is by the use of NMR spectroscopy.

A convenient method to find whether a compound is aromatic, non-aromatic or anti-aromatic is to consider a model of linear chain of sp^2-hybridised atoms that carries the same number of π-electrons as the cyclic compound under consideration. Subsequently, imagine the cyclisation of the linear chain by abstracting two hydrogen atoms from the end of this chain. In case the ring has lower π-electron energy than the open chain, then the ring compound is *aromatic*.

© The Author(s) 2023
V. K. Ahluwalia and R. Aggarwal, *Alicyclic Chemistry*,
https://doi.org/10.1007/978-3-031-36068-8_10

On the other hand, if the ring compound and the open chain have the same π-electron energy, then the ring compound is non-aromatic. Finally, if the ring compound has greater π-electron energy than the open chain compound, then the ring compound is anti-aromatic.

The calculations involved and the experiments used in determining the π-electron energies are beyond our present discussion. However, the following examples illustrate how this approach can be used.

10.2.1 Cyclobutadiene

Consider the following hypothetical transformation:

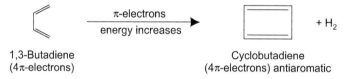

1,3-Butadiene
(4π-electrons)

Cyclobutadiene
(4π-electrons) antiaromatic

Calculations and experiments confirm that π-electron energy of cyclobutadiene is higher than that of its open chain counterpart (1,3 butadiene), so cyclobutadiene is anti-aromatic.

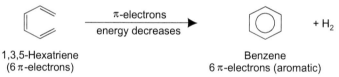

1,3,5-Hexatriene
(6 π-electrons)

Benzene
6 π-electrons (aromatic)

That benzene is aromatic, it is confirmed from the calculations and experiments indicating that benzene has a lower π-electron energy than 1,3,5,-hexatriene.

10.2.2 Cyclopentadienyl Anion

Consider the following hypothetical transformation:

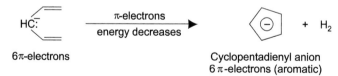

6π-electrons

Cyclopentadienyl anion
6 π-electrons (aromatic)

Cyclopentadienyl anion (on the basis of calculations and experiments) is found to have a lower π-electron energy than its open-chain precursor. So, it is aromatic.

That cyclopentadienyl anion is aromatic can also be predicted on the basis of Huckel's rule.

Cyclopentane is not aromatic as seen by its orbital structure. Also, it does not have the right number of π-electrons, and the π-electrons cannot be delocalised about the entire ring because of the intervening sp^3-hybridised $-CH_2-$ group with no available p orbital. On the other hand, if the $-CH_2-$ carbon atom becomes sp^2 hybridised after it loses a proton (which can be affected by treatment with a strong base to give cyclopentadienyl anion), the two electrons left behind (*see* the representation below) can occupy the new p orbital that is produced.

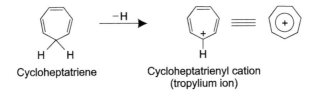

Cyclopentadiene   Cyclopentadienyl anion

Also, the new p orbital can overlap the p orbitals on either side of it to give rise to a ring with six delocalised π-electrons. So all the hydrogen atoms are equivalent (also confirmed by NMR spectroscopy). In case of cyclopentadienyl anion, $4n + 2 = 6$, which is a Huckel number; so, the cyclopentadienyl anion is an aromatic anion. This is also supported by the NMR spectrum of cyclopentadienyl anion which shows a sharp peak in the aromatic region, indicating that all protons are equivalent and hence, the ring is symmetrical.

Similarly, cycloheptatriene (commonly called tropylidene) has six π-electrons, which cannot be fully delocalised because of the presence of the $-CH_2-$ group, a group that does not have an available p orbital (as in the case of cyclopentane).

Cycloheptatriene   Cycloheptatrienyl cation
                          (tropylium ion)

After the removal of a hydride ion from $-CH_2-$ group of cycloheptatriene (by a suitable reagent), a vacant p orbital is created and carbon atom becomes sp^2 hybridised. The resulting cation (cycloheptatrienyl cation) has seven overlapping p orbitals containing six delocalised π-electrons. So, the cycloheptatrienyl cation is an aromatic cation, and all its hydrogen atoms should be equivalent.

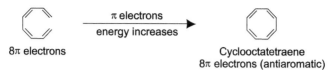

8π electrons   Cyclooctatetraene
                                                8π electrons (antiaromatic)

In this case, calculations and experiments indicate that a planar cyclo-octatetraene would have higher π-electron energy than the openchain octatetraene. So, the planar form of cyclooctatetraene, if it existed, would be antiaromatic.

That cyclo-octatetraene is antiaromatic can also be determined by its procedure described by C.A. Coulson (of Oxford university). In this procedure, inscribe in a circle a regular polygon corresponding to the ring of the compound under consideration (*e.g.*, benzene), so that one corner of the polygon is at the bottom. The points where the corners of polygon touch the circle correspond to the energy levels in the π-orbitals of the system.

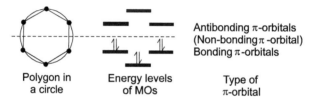

Polygon in   Energy levels   Type of
a circle   of MOs   π-orbital

Antibonding π-orbitals
(Non-bonding π-orbital)
Bonding π-orbitals

Relative energies of the π-molecular orbitals of benzene

With benzene, the above method (called polygon-and-circle method) furnishes the same energy levels as by quantum mechanical calculations. A horizontal line half-way up the circle divides the bonding orbitals from antibonding orbitals. If an orbital falls on this line, it is a non-bonding orbital.

It can now be understood why cyclooctatetraene is not aromatic. It has a total of 8π electrons. Eight is not a Huckel number; it is a 4n number, not a 4n + 2 number. Using the polygon-and-circle method (*see* Fig. below), we find that cyclooctatetraene, if it were planar, would not have a closed shell of π-electrons like benzene. It would have an unpaired electron in each of two non-bonding orbitals. Molecules with unpaired electrons (radicals) are usually not stable; they are highly reactive and unstable. A planar form of cyclooctatetraene should not be like benzene and should not be aromatic.

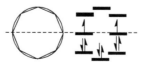

π-molecular orbitals that cyclooctatetraene would have if it were planar

The degeneracy can be removed if the molecule is distorted from maximum molecular symmetry to a structure of lesser symmetry.

It has been shown that cyclooctatetraene exists in a tub structure with alternating single and double bonds. The bonds of cyclooctatetraene are known to be alternately long and short; X-ray studies indicate that they measure 1.48 and 1.34 Å, respectively.

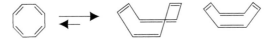

In this tub form, one of the previously degenerate orbital has a lower energy than the other and will be occupied by two electrons In this case, the double bonds are essentially separate and the molecule is still not aromatic. Distortion of symmetry can also occur when one or more carbon atoms are replaced by hetero atoms, or in other ways.

The NMR evidence indicates that the tub form is in quite rapid equilibrium with a very small amount of the planar form at room temperature. The energy difference between the two forms is about 10 kcal. It is, however, interesting to note that the dianion of cyclooctatetraene ($C_8H_6^{2-}$) like azulene, has 10 π-electrons and appears to exist in a planar conformation.

Cyclooctatetraene was first synthesised by Willstätter in 1911 by a thirteen-step route from an alkaloid, pseudopelletierene, isolated from pomegranate bark. Subsequently, it was readily prepared by the polymerisation of acetylene in the presence of nickel cyanide.

N—CH₃ structure (Pseudopelletierene) → 13 steps, 3% overall yield → Cyclooctatetraene ← Ni(CN)₂ / 50°C, 80–90% overall yield ← 4 HC ≡ CH

Pseudopelletierene   Cyclooctatetraene

The chemistry of cyclooctatetraene is very interesting. It undergoes addition reactions with the formation of bridged-ring products such as might be expected on the basis of "valence tautomerism" to bicyclo [4,2,0]-2,4,7-octatriene.

The bridged dichloride on treatment with strong base causes elimination of hydrogen chloride leading to the formation of chlorocyclooctatetraene

Another interesting reaction of cyclooctatetraene is oxidation with mercuric acetate in acetic acid, methanol and water.

10.3 TYPES OF AROMATIC COMPOUNDS

Aromatic compounds are of two types these are benzenoid aromatic compounds and non-benzenoid aromatic compounds.

10.3.1 Benzenoid Aromatic Compounds

Naphthalene, phenanthrene and anthracene are examples of benzenoid aromatic compounds. These are called polycyclic benzenoid aromatic compounds and contain two or more benzene rings fused together.

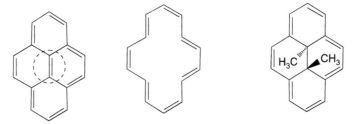

Naphthalene
$C_{10}H_8$

Anthracene
$C_{14}H_{10}$

Phenanthrene
$C_{14}H_{10}$

As seen in anthracene, the two benzene rings are fused together and there is delocalisation of 10 π-electrons over the two rings; thus, the system has a substantial resonance energy. It also shows other aromatic properties like electrophilic substitution.

Anthracene and phenanthrene are isomers; the three rings are fused in a linear way in anthracene and in an angular way in phenanthrene. Both show large resonance energies and other chemical properties characteristic of aromatic compounds.

Pyrene, an interesting aromatic compound, has 16 π-electrons (8 double bonds = 16 π-electrons), sixteen is not a Huckel number. It should be understood that Huckel's rule was introduced originally to be applied only to monocyclic compounds and pyrene is tetracyclic. So, in this case, we look only at the periphery, discarding the internal double bond of pyrene; thus, we have a planar ring with 14 π-electrons. The periphery is very much like that of [14] annulene. Fourteen is a Huckel number (4n + 2, where n = 3), so it can be predicted that the periphery of pyrene would be aromatic by itself in the absence of the internal double bond. The prediction was confirmed by the synthesis of *trans*-15,16-dimethyldihydropyrene (in which the internal double bond was absent), which was shown to be aromatic.

Pyrene (a Kekule structure)   [14] Annulene   *trans*-15,16-Dimethyldihydropyrene

10.3.2   Non-benzenoid Aromatic Compounds

Some examples of non-benzenoid aromatic compounds are the cyclopentadienyl anion, the cycloheptatrienyl cation, *trans*-15,16- dimethyldihydropyrene and the aromatic annulenes (except [6]annulene). Another interesting non-benzenoid aromatic compound is azulene. Let us discuss some of these non-benzenoid aromatic compounds. These include annulenes and azulenes.

10.3.3 Annulenes

Annulenes are conjugated mono-cyclic polyenes, C_nH_n, in which n is 10 or more than 10. The ring size of an annulene is indicated by the number in brackets. A number of annulenes *viz.*, [12], [14], [16], [18], [20], [24] and [30] were prepared by Sondheimer et al. (1959) and tested for Hückel (4n+2) rule. Of these, [14], [18] and [22] annulenes obeying Hückel rule of (4n + 2)π-electron for n=3,4 and 5 respectively, have been found to be aromatic; these annulenes can sustain a diamagnetic ring current. The [16] and [24] annulenes are not aromatic. They are 4n compounds which can sustain a paramagnetic current, not 4n + 2 compounds. The [16] and [24] annulenes had magnetic properties required for its aromatic character, but both these behave chemically like an alkene.

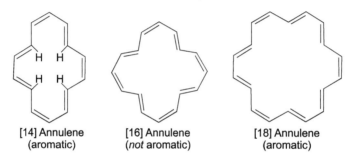

| [14] Annulene | [16] Annulene | [18] Annulene |
| (aromatic) | (*not* aromatic) | (aromatic) |

The aromatic character of [14], [18] and [22] annulenes has also been supported by their NMR spectra. For example, the NMR spectrum of [14] annulene in a fresh solution shows a strong signal at τ4.42 (very close to the value for olefinic protons) and a weak signal at τ3.93. The strong signal at τ 4.42 shows that the [14] annulene does not appear to be aromatic. However, at low temperature (– 60°C), the NMR spectrum shows two signals, one at τ 2.4 and the other at τ 10.0. The former signal (τ 2.4) corresponds to the peripheral protons and the latter to the four inner protons. The NMR behaviour of [14] annulene indicates that it exists in two conformations at room temperature (which have been separated); one being in the solid state and the other as an equilibrium mixture in the solution form. At room temperature, the signals coalesced to form a peak at some intermediate value. The [14] annulene is not completely planar owing to the transannular interactions of the four inner hydrogen atoms. However, the molecule has a completely cyclic conjugated system and is aromatic.

Dehydro [14] annulene, $C_{14}H_{12}$, is a (4n) molecule though it shows the magnetic properties required for aromaticity. In this molecule, the two inner hydrogen atoms are too apart to experience transannular interaction, and so the molecule is planar. 1,8-Bis dehydro [14] annulene, $C_{14}H_{10}$, is a (4n + 2) molecule and has magnetic property. However, it is not possible to represent this molecule with conjugated double bonds.

A number of bridged[14] annulene have been synthesised and shown to be aromatic. X-ray diffraction studies show that the bridged [14] annulenes ring are not quite planar with single bond length alteration and thus have weaker ring current.

Dehydro [14] annulene   1,8-Bis-dehydro [14] annulene

[10] annulene is the smallest annulene of the (4n+2) type. It was prepared by the photolysis of *trans*–9,10-dihydronaphthalene (Van Tamelen, et al., 1971). It can be represented by the following three structures (A, B and C).

The [10] annulene (A) has two *trans* double bonds; the bond angles are approximately 120° and so have no appreciable angle strain. Because of the two hydrogen atoms in the centre of the ring interfering with each other, the ring is prevented from becoming coplanar. Since the ring is not planar, the *p* orbitals of the carbon atoms are not parallel and so cannot overlap effectively around the ring to form the π molecular orbitals of an aromatic system.

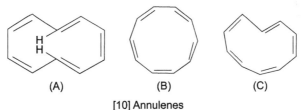

(A)   (B)   (C)

[10] Annulenes
None is aromatic because none is planar

The representation (B) of [10] annulene with all *cis* double bonds, if it were planar, has considerable angle strain because of the internal bond angle of 144°. So, any stability this isomer had by becoming planar, would be offset by the destabilising effect of the increased angle strain in order to become aromatic. Similarly, large angle strain associated with a planar form inhibits molecule of [10] annulene isomer with one trans double bond (C) from becoming aromatic.

10.3.4 [18] Annulene

The interesting large difference in NMR chemical shifts between the inside and outside hydrogens of annulenes (which in fact supports their aromatic character), as in the case of [14] annulene, has already been mentioned. In case of [18] annulene, the 12 outer protons are found at about δ 9 and the 6 inner protons at about δ –3. X-ray crystallography shows that it is nearly planar, so that interference of the inner hydrogens is not important in this annulene. It is reasonably stable, undergoes aromatic substitution. The C – C bond distances are not equal, but they do not alternate. There are 12 inner bonds of about 1.38 Å and 6 outer bonds of about 1.42 Å. It has a resonance energy of about 37 kcal/mol, similar to that of benzene.

[18] Annulene

[18]Annulene can be synthesised by the oxidative coupling of 1,5-hexadiyene to give 6% yield of a trimer, which rearranges in the presence of potassium *tert*. butoxide to the fully conjugated 1,2,7,8,13,14-tridehydro [18] annulene. Final hydrogenation over lead-poisoned palladium on calcium carbonate catalyst (Lindlar catalyst) gives [18] annulene as a brown-red crystalline solid which is reasonably stable in the presence of oxygen and light.

$HC \equiv C — CH_2 — CH_2 — C \equiv CH$

1,5-Hexadiyne

$\xrightarrow[C_5H_5N]{Cu^{II}}$

Trimer

$\xrightarrow{K^+\bar{O}C(CH_3)_3}$

$\xrightarrow[Pd\,(Pb)]{H_2}$

[18] annulene

10.3.5 Metallocenes

These are organometallic compounds having 'sandwich' structure. In these, iron or any other transition metal like Cr, Mn, Ni, Ti, Ru, Os, form a π complex with unsaturated organic compounds like cyclopentadiene, and benzene, etc. One of the most interesting and important example of metallocene is ferrocene.

Ferrocene

It is obtained by the reaction of Grignard reagent of cyclopentadiene with ferrous chloride, the Grignard reagent in turn is obtained by the reaction of cyclopentadiene with phenyl-magnesium bromide. The overall yield is 71% from cyclopentadiene.

| Cyclopentadiene | Phenyl-magnesium bromide | | Cyclopentadienyl magnesium bromide | Benzene |

$$2 \left[\ominus \right] Mg^{2+}Br^- + FeCl_2 \longrightarrow (C_5H_5)_2Fe + 2Mg\,BrCl$$

Ferrocene

Ferrocene is an orange solid with a melting point of 174°C. It is a highly stable compound and can be sublimed at 100°C and is not damaged when heated upto 400°C. Its structure has been established beyond doubt by X-ray studies as '*sandwich*' structure in which iron is bound to two parallel cyclopentadienyl rings.

Ferrocene

Ferrocene has zero dipole moment and so the molecule is symmetrical. Also, the infrared spectrum shows that all the C – H bonds are equivalent. In ferrocene and other cyclopentadienyls of transition metals (Cr, Mn, Co, etc.), the entire ring is bonded uniformly to the metal atom; the bonding occurs by overlap of the sextet of π electrons of the rings with the *d* orbitals of the metal, thereby giving a delocalised covalent bond between the metal atom and the cyclopentadienyl ring.

Ferrocene is quite resistant to the action of acids and bases, even to concentrated sulphuric acid. However, it is readily oxidised with nitric acid to the less stable ferricinium ion.

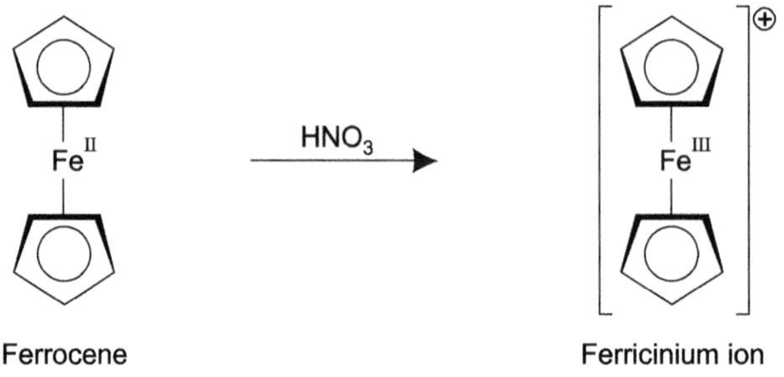

Ferrocene   Ferricinium ion

Ferrocene is an aromatic compound like benzene, and does not react easily by addition but enters readily into electrophilic substitution reactions. Thus, Friedel-Crafts acylation gives monoacetyl ferrocene, but in the presence of excess aluminum chloride, a diacetyl ferrocene is obtained. The second acetyl group enters the second ring, and because only one such diacetyl derivative can be obtained, the two cyclopentadienyl rings are most likely free to rotate about the bond axis to the metal. In fact, the free rotation about the axis has also been shown by a number of other studies. The bonding is such that the rings of ferrocene are capable of essentially free rotation about an axis that passes through the iron atom that is perpendicular to the rings.

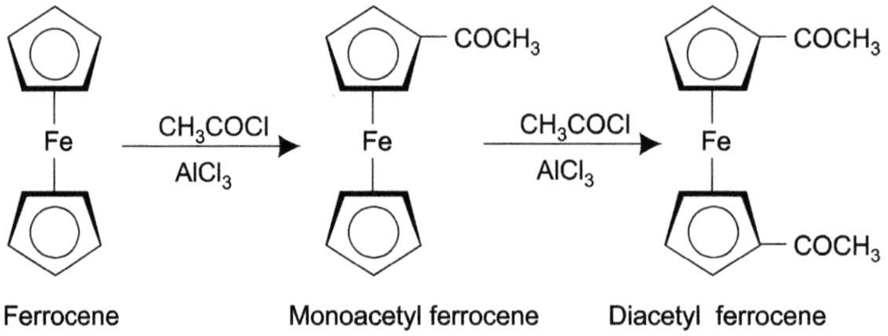

Ferrocene   Monoacetyl ferrocene   Diacetyl ferrocene

The iron of ferrocene has 18 valence electrons and is, therefore, coordinatively saturated. The valence electrons are calculated as follows:

Iron has 8 valence electrons in the elemental state and the oxidation state of iron in ferrocene is +2. Therefore $d^n = 6$.

$$d^n = 8 - 2 = 6$$

Each cyclopentadienyl (cp) of ferrocene donates 6 electrons to the iron. Therefore, for the iron, the valence electron count is 18.

Total number of valence electrons $= d^n + 2(cp)$

$$= 6 + 2(6)$$

$$= 18$$

'Half-sandwich' compounds have also been prepared using metal carbonyls. Some of these are shown below:

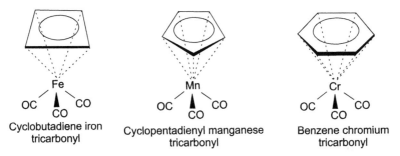

| Cyclobutadiene iron tricarbonyl | Cyclopentadienyl manganese tricarbonyl | Benzene chromium tricarbonyl |

Ernsto, Fischer and Geoffrey Walkinson received the Nobel Prize in 1973 for their pioneering work on the chemistry of metallocenes.

Bis-benzene chromium : It is analogous to ferrocene, as it is the complex of two benzene molecules with chromium. The bonding involves the π-electrons of the two benzene rings.

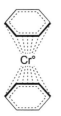

Bis-benzene chromium

Bis-benzene chromium is thermally quite stable, but is less stable than ferrocene and melts with decomposition at about 283°C into benzene and chromium. Also, it lacks aromatic character of either benzene or ferrocene. In fact, no electrophilic substitution reactions have been reported.

Transition metal complexes of cyclobutadiene: Cyclobutadiene has been predicted to be an unstable system as the number of π-electrons (four) is energetically unfavourable. This prediction has been supported by experimental data. Thus, tetramethyldichlorocyclobutene on treatment with lithium amalgam in the presence of nickel carbonyl gives a nickel chloride complex of tetramethylcyclobutadiene whose structure has been established by X-ray data.

In this case, the π-electron system of cyclobutadiene derivative is stabilised by complex formation. In the above reaction, if nickel carbonyl is not added, a dimer of tetramethylcyclobutadiene is obtained *via* an intermediate dechlorinated product.

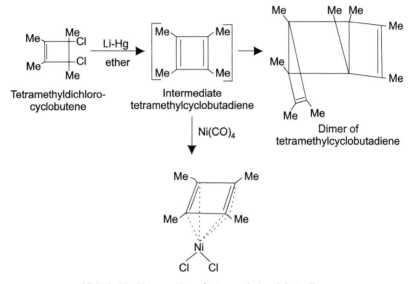

Nickel chloride complex of tetramethylcyclobutadiene

10.3.6 Azulenes

Azulenes are bicyclic compounds containing a seven-membered ring fused to a five-membered ring through adjacent carbon atoms. These were obtained from the oils of camomile, yarrow and cubeb, which contain sesquiterpenoids. The sesquiterpenoids when subjected to operations such as distillation at atmospheric pressure, treatment with acids or oxidising agents, steam distillation, which result in dehydrogenation, gave azulenes. One such derivative of azulenes is guaiol, which occurs in guaiacum wood oil. It was given the structure (I) on the basis of a large number of studies.

(I)
Guaiol

The name azulene is given to the parent compound of the azulene series, $C_{10}H_8$.

The following numbering is used.

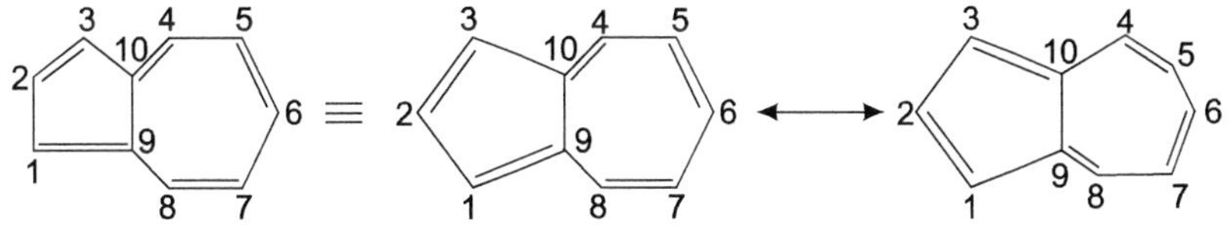

The fully saturated (or hydrogenated) azulene has been given the name bicyclo[5,3,0]decane (Baeyer); in this case the name is derived by numbering the carbon atoms as shown above.

The parent compound, azulene, is also called cyclopentacy-cloheptene or bicyclo[5,3,0]deca-1,3,5,7,9-pentaene.

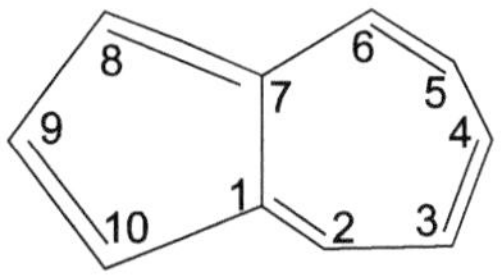

Azulenes are compounds that possess some measure of aromatic character (Resonance energy = 205 kJ/mol.). So these are classified as non-benzenoid aromatic compounds.

One of the most striking properties of azulenes is their intense blue or blue-violet colour, noticeable even at very high dilution. Azulene has significant dipole moment of 1.0 D with a five-membered ring at the negative end of the dipole. The structure can be represented as a hybrid of neutral and ionic molecules.

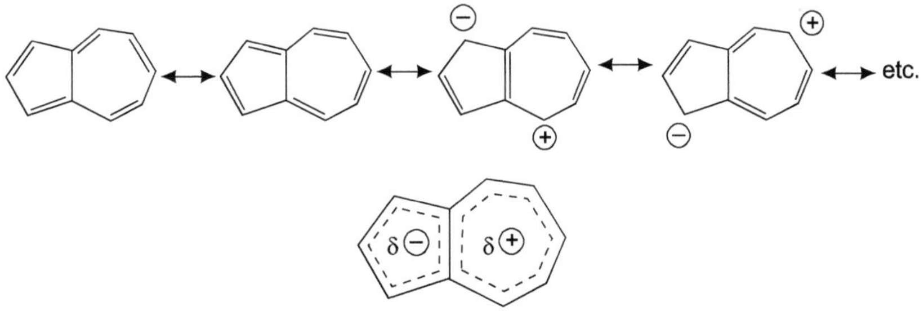

Resonating and hybrid structures of Azulene

10.3.7 Synthesis of Azulenes

1. The parent compound, azulene, $C_{10}H_8$, was obtained in small amounts by the dry distillation of calcium adipate. Following steps were proposed (Plattner and Pfau, 1930).

Azulene itself has been isolated from tobacco smoke (Ikeda, 1947) and also from caucal oil (Mitui, 1941).

2. The first total synthesis of azulene was carried out from cyclopenteno cycloheptanone made from 1,6-cyclodecadione (Pfau and Plattner, 1936).

The above method is also useful for the synthesis of some 4-alkyl azulenes starting from the intermediate cyclopenteno cycloheptanone.

The above synthesis also confirms the structure of azulene.

3. Diazoacetic ester method

This is the most widely used method for the synthesis of azulene. It consists in the reaction of indane (or its derivatives) with ethyl diazoacetate (Plattner, 1940). The reaction involves ring expansion. In this procedure, the reagent may react with the various Kekule forms of indane. However, the end product is the same.

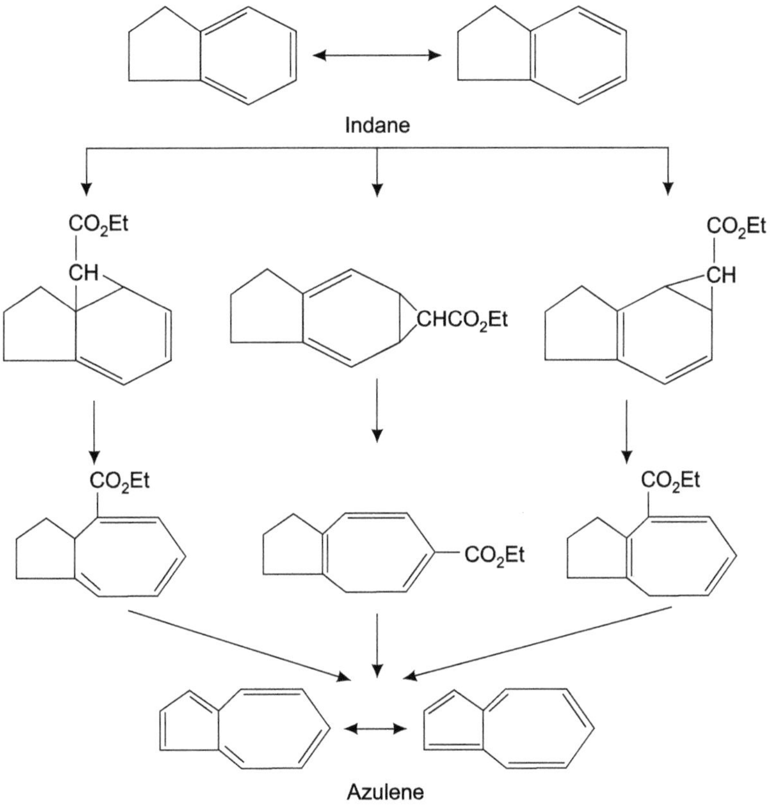

Azulene

In the above method, the formation of three isomeric carboxylic esters is shown. However, it shows that in some substituted indanes only two products may be obtained. These can be separated by chromatography on alumina The pure esters are dehydrogenated (Pd–C) to give azulene carboxylic esters, which can be converted into free acids. Thus, this method can also be used for the synthesis of azulene carboxylic acids, which in turn can be converted into other esters, amide, etc.

4. Demjanow ring expansion

In this reaction, an appropriate amino compound is treated with nitrous acid.

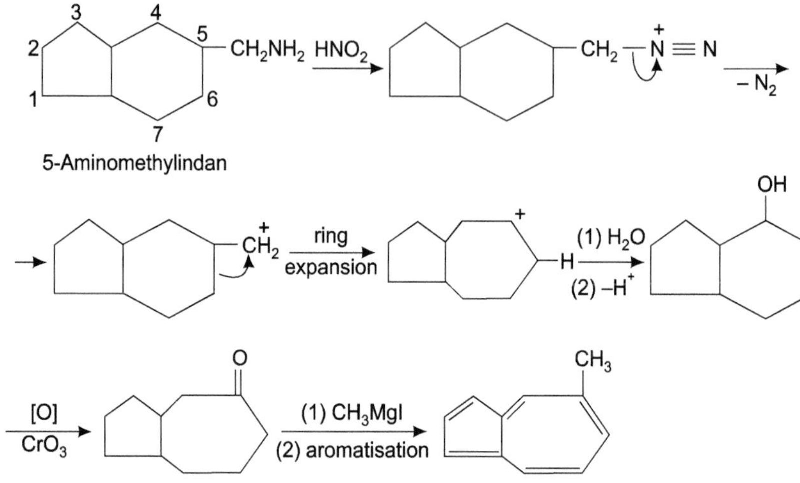

5-Aminomethylindan

5-methylazulene

5. Azulene from cycloheptanone

2-Alkylazulenes are prepared from cycloheptanone as follows: (Plattner, 1946).

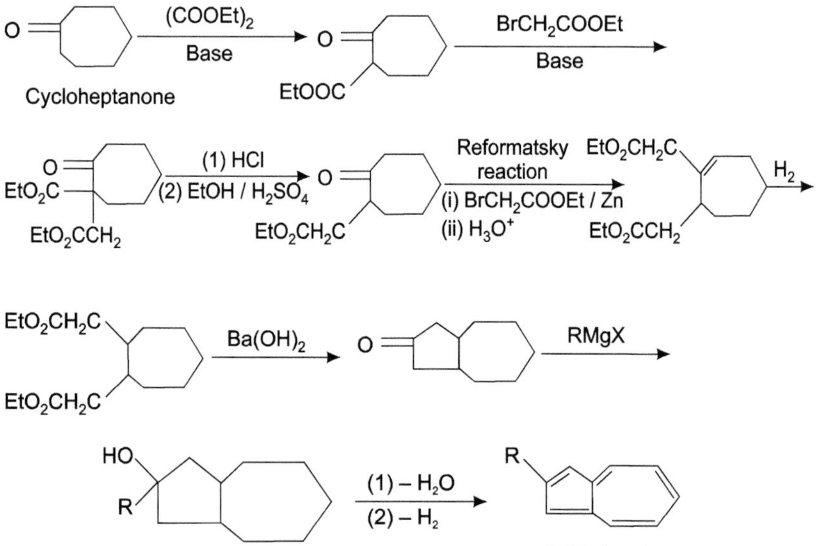

Cycloheptanone

2-Alkylazulene

Properties : Azulene (which is isomeric with naphthalene) is less stable than naphthalene and isomerises quantitatively on healing above 350°C in the absence of air.

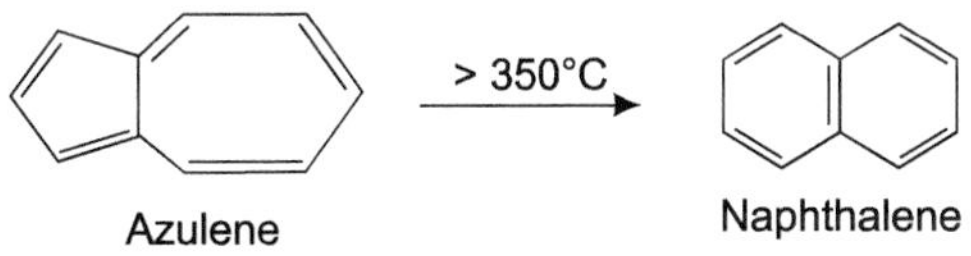

Azulene   > 350°C   Naphthalene

Chemically, azulene behaves as an aromatic compound, *e.g.*, it undergoes bromination, nitration, etc. Electrophilic substitution occurs preferentially in the five-membered ring, since this is more electron rich than the seven-membered ring. Thus, substitution takes place at positions 1- and 3-. For example, Friedel-Crafts acetylation leads to a mixture of 1-acetylazulene and 1,3-diacetylazulene.

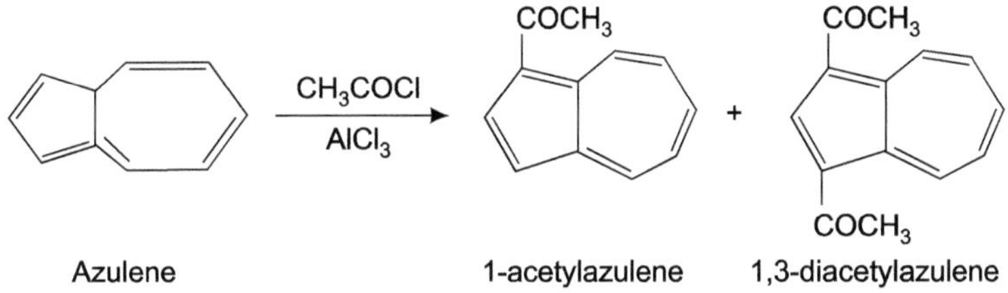

Azulene CH_3COCl / $AlCl_3$ 1-acetylazulene   +   1,3-diacetylazulene

In the presence of strong acids, the 1-position is protonated to give a derivative of the relatively stable cycloheptatrienyl (tropylium) ion.

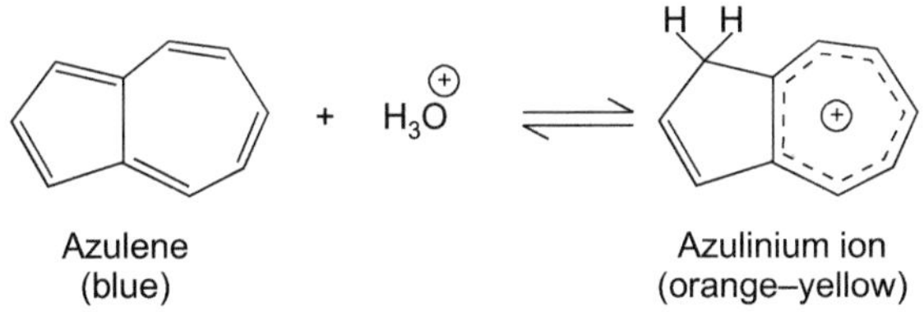

Azulene
(blue)

+ $H_3O^{\oplus}$ ⇌

Azulinium ion
(orange–yellow)

Bridged Rings

11.1 INTRODUCTION

Bridged ring compounds contain two or more rings fused together. A study of these compounds is of special interest in relation to ring strain and their characteristic properties. Some of the typical examples are discussed below. These compounds are named as derivatives of butane, pentane, hexane, heptane, octane, etc., depending on the total number of carbon atoms in their rings combined together. The number of carbon atoms on the bridge head carbon atoms are indicated by the number 0 (if there is no carbon atom in between the bridge head), 1 (if there is one carbon atom in between the bridge head), 2 (if there are two carbon atoms between the bridge head) and so on. The following examples illustrate the nomenclature of bridged rings.

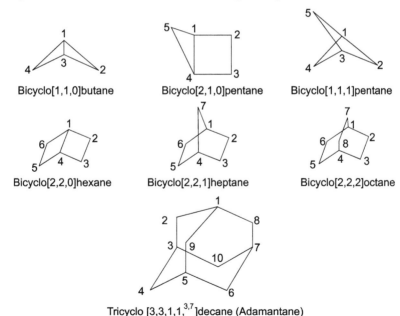

Bicyclo[1,1,0]butane   Bicyclo[2,1,0]pentane   Bicyclo[1,1,1]pentane

Bicyclo[2,2,0]hexane   Bicyclo[2,2,1]heptane   Bicyclo[2,2,2]octane

Tricyclo [3,3,1,1,^{3,7}]decane (Adamantane)

© The Author(s) 2023
V. K. Ahluwalia and R. Aggarwal, *Alicyclic Chemistry*,
https://doi.org/10.1007/978-3-031-36068-8_11

11.2 BICYCLO[1,1,0]BUTANE

It is obtained by the condensation of diethyl malonate with 1,3-dibromopropane.

1,3-Disubstituted bicyclo[1,1,0]butane can be synthesised by the following cycloaddition reaction.

The central bond length of bicyclo[1,1,0]butane has been found to be 1.49Å which correlates with the observed microwave spectrum.

Bicyclobutane is decomposed on heating to give butadiene; the reaction takes place *via* a radical intermediate.

The thermal decomposition is found to be stereospecific as shown below.

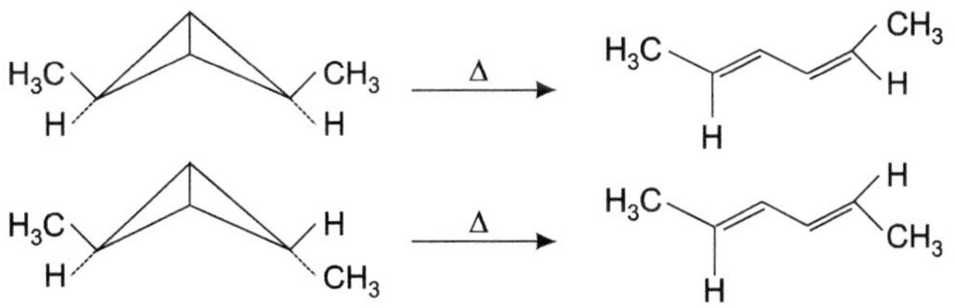

In view of the above, the thermal decomposition of bicyclobutane takes place in a concerted pathway.

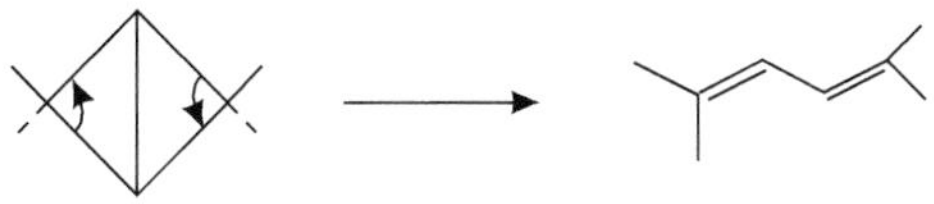

Acidic fission of bicyclobutane gives cyclobutanol along with cyclopropane derivative in equal amounts.

| Bicyclobutane | | Cyclobutanol | | Cyclopropane derivative |

1-methyl-cyano bicyclo[1,1,0]butane undergoes ring fission under alkaline conditions. Thus,

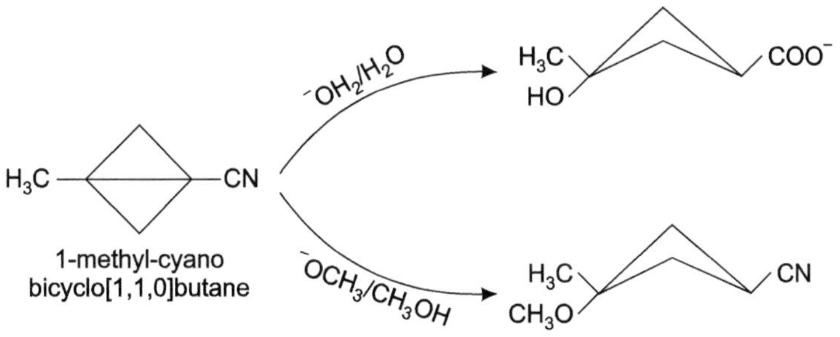

11.3   BICYCLO[1,1,1]PENTANE AND BICYCLO[2,1,1] HEXANE

Bicyclo[1,1,1,]pentane is the smallest bridged ring system possible that has been synthesised. The two bridged carbons are in close proximity (ca. 200 pm) because of very small endocyclic valency angle (72.5°). It is unstable and is converted into 1,4-pentadiene on heating to 300°C. The next higher homologue is bicyclo[2,1,1]hexane with the smallest endocyclic angle of 81.2°.

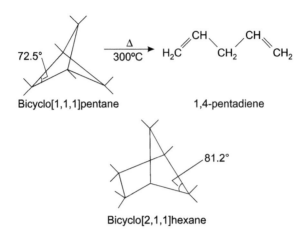

Bicyclo[1,1,1]pentane   1,4-pentadiene

Bicyclo[2,1,1]hexane

11.4   BICYCLO[2,1,0]PENTANE

It is obtained from cyclopentadiene by reacting either with azodicarboxylic ester or with nitrosobenzene as given below:

(a)   Cyclopentadiene + azo compound → Azodicarboxylic ester

Cyclopentadiene   Azodicarboxylic ester

H_2 catal. → ... $\overline{O}H$ / H_2O →

NH/NH $\xrightarrow[\text{oxidation}]{Cu^{2+}}$ N=N $\xrightarrow[-N_2]{160°C}$ ⬡ = Bicyclo[2,1,0]pentane

(b)   Cyclopentadiene + PhNO → N–Ph / O $\xrightarrow{H_2\ \text{catal.}}$

Cyclopentadiene   Nitrosobanzene

N–Ph / O $\xrightarrow{200°C}$ ⬡ + ⬠ + PhNO

Electron diffraction studies of the molecular structure of bicyclo[2,1,0] pentane in the vapour phase shows that the cyclobutane ring is planar and the dihedral angle between the rings is 109.4°±0.4°. The zero-membered bridge bond is short (1.439Å) and the C – C bond opposite in the cyclobutane ring is long (1.622Å).

Unlike bicyclo[1,1,0]butane, bicyclo[2,1,0]pentane is comparatively thermally stable. However, at 350°C, it is isomerised to cyclopentene.

It also reacts rapidly with hydrogen bromide to give cyclopentyl-bromide.

Cycloaddition with maleic anhydride occurs on the "inside" face of the molecule.

11.5 BICYCLO[2,2,0]HEXANE

It is obtained from cyclohexadiene *via* a nitrosobenzene adduct.

Bicyclo[2,2,0]hexane (60%)

Bicyclo[2,2,0]hexane can be easily isomerised to hexa-1,5-diene, which is a by-product in the above synthesis.

The NMR spectra of bicyclo[2,2,0]hexane gives a broad signal (corresponding to 8 protons) centered around τ 7.60 compared to a signal at τ 8.04 for cyclobutane.

11.6   BICYCLO [2,2,0] HEXADIENE

It is obtained by photocyclisation of cyclohexadiene dicarboxylic anhydride.

The above sequence of reactions is also used for the synthesis of bicyclo[2,2,0] hexene.

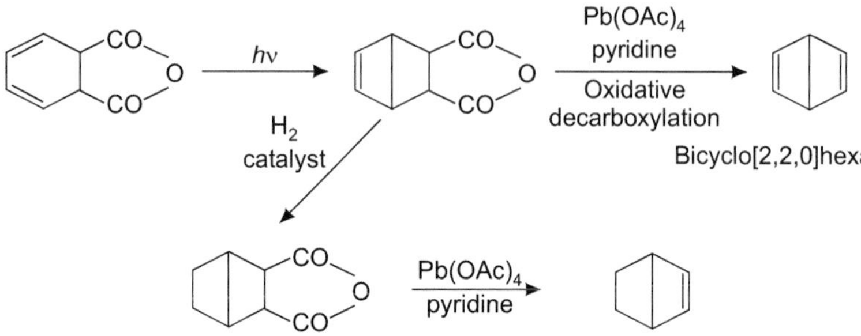

Bicyclo[2,2,0]hexadiene

Bicyclo[2,2,0]hexene

Inspite of considerable ring strain, bicyclo[2,2,0]hexadiene is quite stable. In fact, it represents 'Dewar' benzene and on keeping reverts back to benzene; this conversion requires considerable activation energy.

The hexamethyl substituted bicyclo[2,2,0]hexadiene is obtained along with hexamethyl benzene by aluminium chloride catalysed trimerisation of but-2-yne.

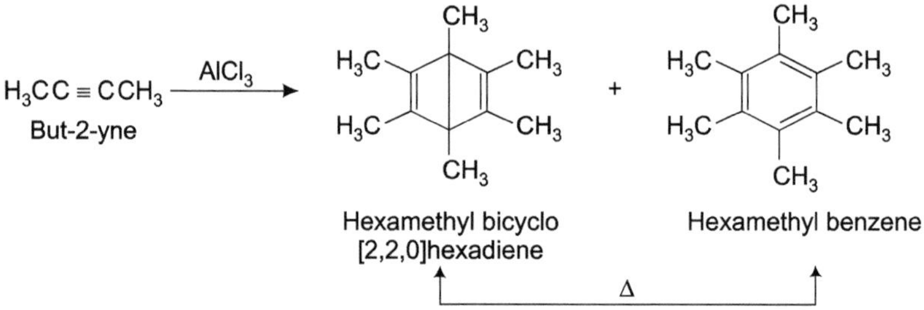

Hexamethyl bicyclo
[2,2,0]hexadiene

Hexamethyl benzene

11.7   BICYCLO[2,2,1]HEPTANE

It is commonly known as norbornane. In the structure of norbornane, cyclohexane ring has the boat conformation and has two cyclopentane units (1,2,3,4,7 and 4,5,6,1,7) bent along the 1,4-axis. The substituents at positions 2,3,5 and 6 may be *endo-* or *exo-* oriented as shown here:

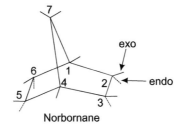

Norbornane

The degree of bond deformation in the structure of bicyclo[2,2,1] heptane is indicated by the bond angles [1,7,4 = 93°; 1,2,3 =105°; 6,1,2 = 114° and 7,1,2 = 98°], as determined by X-ray data and also the infrared absorption of the carbonyl group at position 2 (v_{co} 1751 cm^{-1}) and position 7 (v_{co} 1773 cm^{-1}). In addition, strain is introduced by the 1,4-bridge.

Synthesis of bicyclo[2,2,1]heptane and its derivatives.

Exo-norbornylacetate is obtained in good yield by acetolysis of cyclopent-3-enyl ethyl toluene-p-sulphonate.

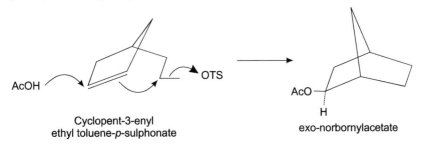

AcOH

OTS

AcO

H

Cyclopent-3-enyl
ethyl toluene-p-sulphonate

exo-norbornylacetate

A more generally useful method for the synthesis of norbornane derivatives consists in the Diels-Alder addition of different reagents to cyclopentadiene (there is predominantly endo addition in the Diels-Alder addition).

Cyclopentadiene

X = H, OAc, CO₂Me

endo

exo

X = CO₂Me, endo: exo = 76:24

Synthesis of norbornane derivative

Some other derivatives of bicyclo[2,2,1]heptane are obtained from bicyclo[2,2,1]heptene.

11.8   ADAMANTANE

It is a tricyclic system having three cyclohexane rings. Its structure is similar to hexamethylene tetramine (which is obtained by the reaction of formaldehyde and ammonia), in which the four nitrogens are replaced by CH.

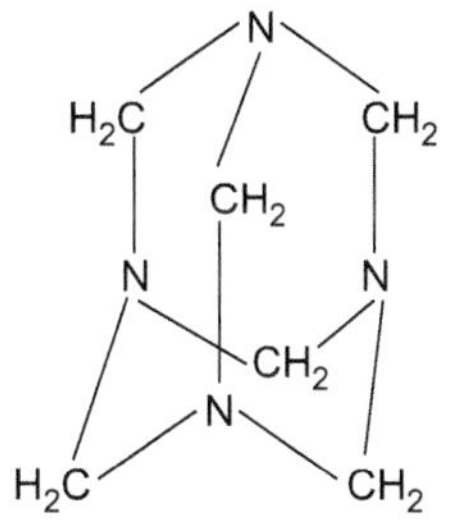

Hexamethylenetetramine
($C_6H_{12}N_4$)

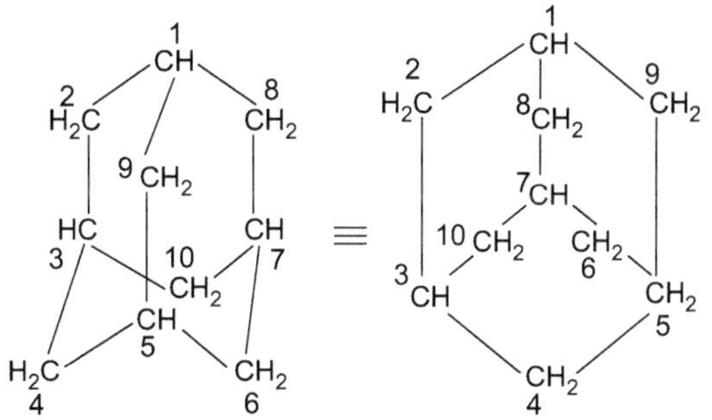

Adamantane ($C_{10}H_{16}$)

Tricyclo[3,3.1.1^{3,7}]decane

Adamantane is named tricyclo[3,3,1,1^{3,7}]decane. In numbering these polycyclic ring systems, a junction is chosen as C_1. The numbering then proceeds along the longest linkage to the next junction, continues along the next longest path and is finally completed along the shortest path. In adamantane, the largest ring and its main linkage form a bicyclic system, and the location of the fourth or secondary linkage is shown by superscripts.

Adamantane has a symmetrical and strainless structure. It is obtained from the dimer of dicyclopentadiene by hydrogenation followed by the action of strong acids like HBF_4, $HAlCl_4$, FSO_3H or Lewis acid such as $AlCl_3$.

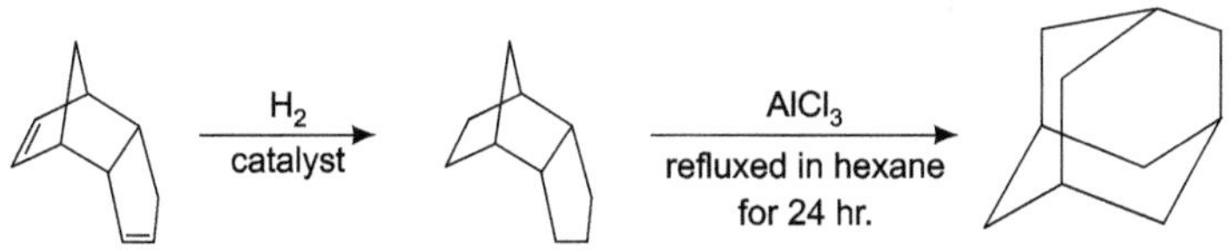

dimer of cyclopentadiene

Adamantane yield
91%, no intermediate isolated

Proposed Machanism

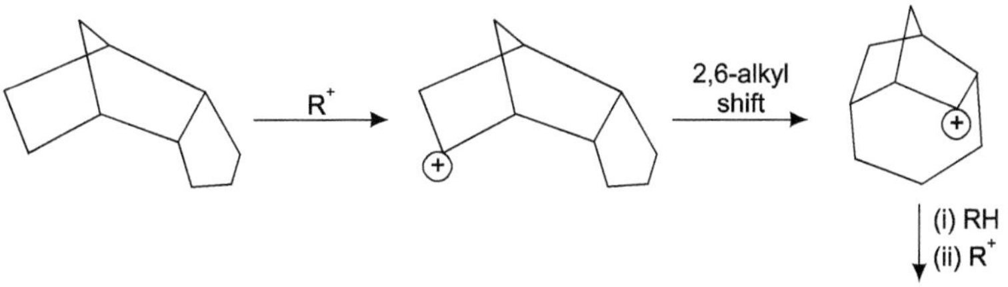

Ref. J.A.C.S., 79, 3292, (1957)
J.A.C.S., 82, 4645 (1960)

Another synthesis of adamantane involves the reaction of formaldehyde with malonic ester following the steps given below:

□ represents C of CH$_2$O
Four circles represent 4 molecules of dimethyl malonate ester

(1) W-K Redn.
NH₂NH₂/C₂H₅ONa
(2) Hydrolysis COOMe → COOH
(3) Ag salt/Br₂
   Hunsdiecker reaction
   COOH → COOAg → Br

Dibromo adamantane   Adamantane

Here, –COOMe group is removed because structure (1) becomes a bicarbonyl system attached to the carbon marked with asterisk (C*). This becomes *tert.* and prevents change into enol form; therefore one –COOMe is removed. This molecule is made up of 3 moles of formaldehyde and 4 moles of malonic ester.

Although the model of adamantane appears to be strain free, the molecule has about 76 kcal/mole strain. This is more than four time the strain in cyclohexane. In adamantane, the framework structure contains the bond angles close to tetrahedral value. However, the $-C-CH_2-C-$ bond angle is not exactly tetrahedral, the optimum value being 112.5° and in slightly strained cyclohexane, the angle is 111.5°. This causes strain in adamantane.

In the presence of aluminium halides, 2-methyladamantane is converted almost completely into 1-methyl derivative. This rearrangement proceeds *via* the formation of adamantyl cation.

Adamantane undergoes bridge head substitution *via* the adamantyl ion.

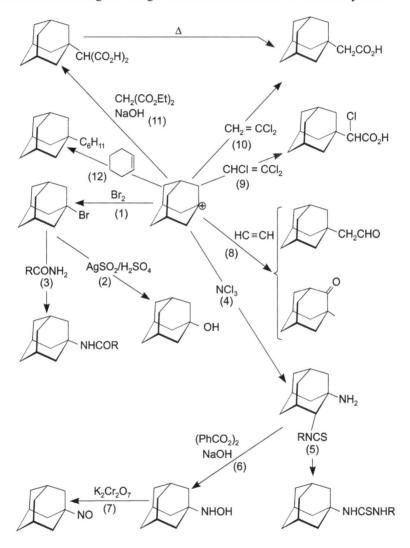

Bridge-head substitutions in Adamantane

The adamantyl-1-cation is a structure of some stability. It has been prepared in solution, *e.g.*, from adamantyl fluoride and SbF$_5$, and characterised by its NMR spectrum, which interestingly shows greater deshielding at the β-carbon than the α-carbon from the cationic centre.

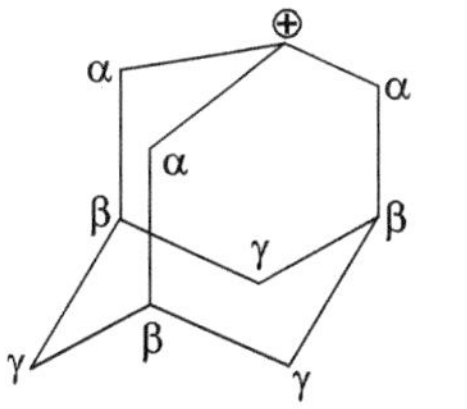

α 6H   τ 5.50
β 3H   τ 4.58
γ 6H   τ 7.33

11.9 TWISTANE

The structure of twistane is isomeric with adamantane. It has been synthesised by the following sequence of reactions (Whitelock, 1962).

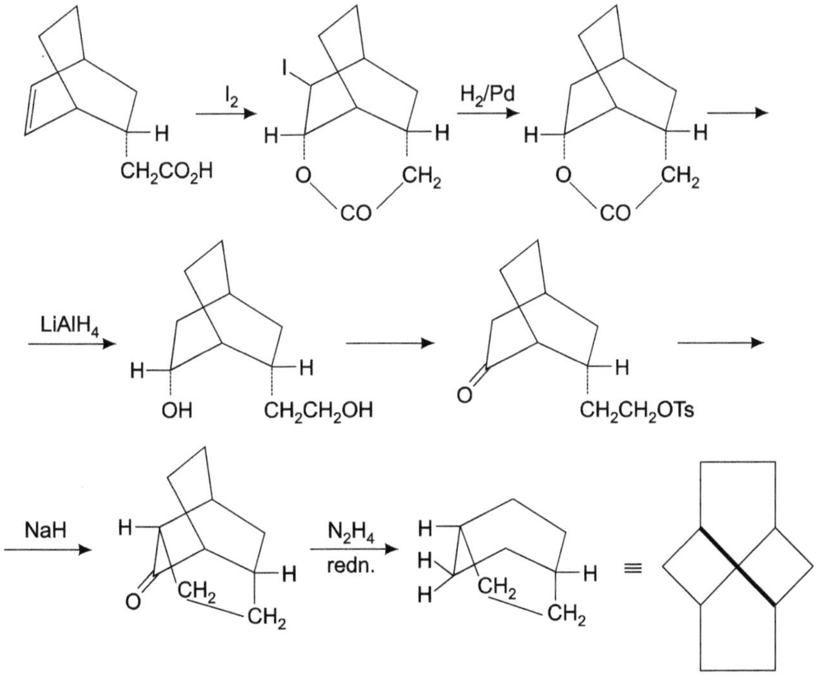

Twistane

Twistane is a disymmetric structure. (+)-Twistane [[α]$_D$ 414°], was obtained starting from the optically active carboxylic acid.

(±) Twistane

The Cage Molecules

12.1 INTRODUCTION

Certain polycyclic hydrocarbons exhibiting fascinating, cage-like structures have been synthesised. These are generally given names indicating their shape.

12.2 CUBANE

It was synthesised (Pettit et al., 1966) by the decomposition of cyclobutadiene iron tricarbonyl complex in the presence of 2,5-dibromo-1,4-benzoquinone. The cyclobutadiene, which is released spontaneously by the decomposition of the iron tricarbonyl complex, gives an adduct (an endo isomer as obtained in Diels-Alder addition). Photochemical cyclisation of the adduct gave dibromodiketone, which was converted into cubane dicarboxylic acid by **Favorskii rearrangement**.

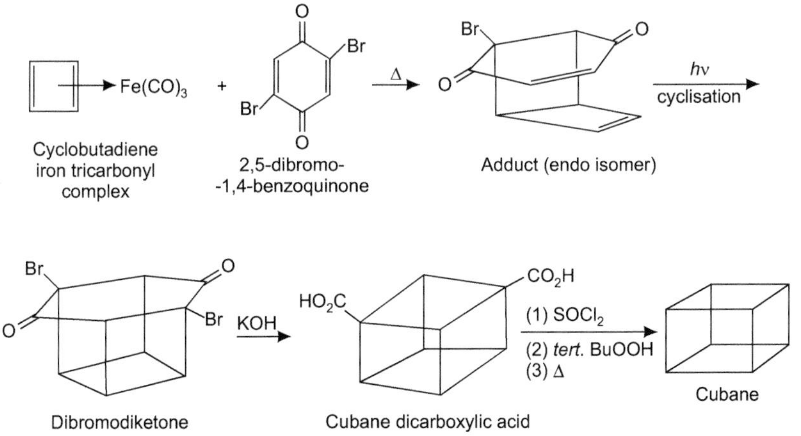

Cyclobutadiene iron tricarbonyl complex

2,5-dibromo--1,4-benzoquinone

Adduct (endo isomer)

Dibromodiketone

Cubane dicarboxylic acid

Cubane

© The Author(s) 2023
V. K. Ahluwalia and R. Aggarwal, *Alicyclic Chemistry*,
https://doi.org/10.1007/978-3-031-36068-8_12

The final step of decarboxylation was affected by using the thermal decomposition of the *tert*.butyl perester by heating in isopropyl benzene.

$$\equiv\!C - COCl + HOOCMe_3 \longrightarrow \equiv\!CO - O - O - CMe_3 \longrightarrow$$
$$\equiv\!\overset{\bullet}{C} + CO_2 + Me_3CO^{\bullet}$$

$$\equiv\!\overset{\bullet}{C} + RH \longrightarrow \equiv\!C - H + \overset{\bullet}{R}$$
(RH = solvent)

Cubane, as expected, shows a single NMR proton signal at δ 4.04 and a single 13_C resonance at 47.3. The structure as determined by electron diffraction studies shows a bond length of 157.5 pm (1.57 Å), larger than that in cyclobutane. Earlier X-ray data indicated a C – C – H bond angle of 123 – 127°, suggesting a high degree of *s*-character of the C – H bond. Surprisingly, J_{13C-H} of 60 Hz suggested the high *s*-character of the C – H bond.

12.3 PRISMANE

It represents a prismatic structure, once proposed for benzene by Ladenburg. In fact, prismane and benzene are called valence isomers.

Despite a strain energy of at least 300 kJ/mole, the prismane structure is remarkably stable (Oth, 1968; Woodward et al., 1970). Prismane has been synthesised as follows:

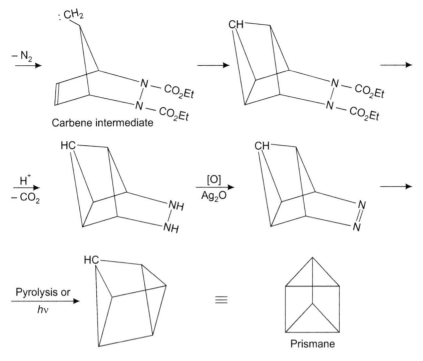

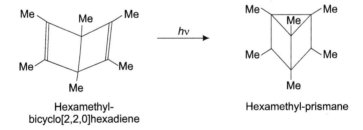

Various alkyl derivatives of prismanes are known and are obtained by the photochemical cyclisation of a bicyclo [2,2,0] hexadiene. Thus, hexamethyl prismane is obtained from hexamethyl bicyclo [2,2,0,] hexadiene (Schäfer et al., 1967).

Hexamethyl-bicyclo[2,2,0]hexadiene	Hexamethyl-prismane

The starting compound, *viz.*, hexamethyl bicyclohexadiene, is obtained by cyclotrimerisation of but-3-yne.

Hexamethyl prismane on heating isomerises to reform hexamethyl bicyclo [2,2,0] hexadiene along with hexamethyl benzene in the ratio of 2.5 : 1.

Hexamethyl prismane   Hexamethyl   Hexamethyl benzene
bicyclo[2,2,0]hexadiene

The synthesis of hexamethyl prismane by photocyclisation is an allowed process in terms of the correlation of orbital symmetries. However, the symmetry correlation for the thermal retrogression leads to occupation of an anti-bonding level in the products. Thus, it is easy to understand the thermal stability of prismanes, and the high energy of activation of about 130 kJ/mole observed for the thermal decomposition of hexamethyl prismane to hexamethyl benzene and hexamethyl bicyclo[2,2,0]hexadiene.

☐☐☐

Both tropone (cycloheptatrienone) and tropolone (cycloheptatri-enolone) are considered to be derivatives of cycloheptatriene.

Cycloheptatriene   Tropone   Tropolone

Both tropone and tropolone are interesting compounds because they possess aromatic character. Although both contain a keto group, they lack ketonic properties. Tropone may be prepared from anisole by the following sequence of reactions (Birch et al., 1962).

Anisole   Tropone

The tropolones, a class of non-benzenoid aromatic compounds were first encountered in different kinds of natural products. Two such examples are β-thujaplicin (4-isopropyltropolone), isolated from the oil of Formosan cedar, and colchicine, isolated from autumn crocus. The latter is used in medicine for the treatment of gout. It has an effect on cell division and is used in plant genetic studies to cause doubling of chromosomes.

© The Author(s) 2023
V. K. Ahluwalia and R. Aggarwal, *Alicyclic Chemistry*,
https://doi.org/10.1007/978-3-031-36068-8_13

β-Thujaplicin   Colchicine

Tropolone itself is prepared in a number of ways. The most convenient method involves oxidation of 1,3,5-cycloheptatriene (tropilidene) with alkaline potassium permanganate. The yield is low and the product is isolated as copper salt.

Another method involves the reaction of cyclopentadiene and tetrafluoro ethylene.

Tropilidene required for the synthesis of tropolone is obtained by the light induced reaction of diazomethane with benzene or better by the thermal rearrangement of the Diels-Alder adduct of cyclopentadiene and acetylene.

Benzene   Diazomethane   Tropilidene

Cyclopentadiene   Acetylene   Adduct

Tropolone is an acid with an ionisation constant of 10^{-7}, which is intermediate between that of acetic acid and phenol. Like phenols, tropolone forms coloured complexes with ferric chloride solution. The aromatic character of tropolone is shown by its properties. Thus, like benzene, it resists hydrogenation, undergoes diazo coupling and can be nitrated, sulphonated and halogenated. The aromaticity of tropolones can be attributed to resonance involving the two non-equivalent structures (a) and (b) and several structures such as (c) and (d) which correspond to the stable tropylium cation with six π electrons.

(a)   (b)   (c)   (c)

The tropylium cation itself can be easily prepared by the transfer of hydride ion from tropilidene to triphenylmethyl carbocation salts in sulphur dioxide.

Tropilidene   Triphenylmethyl   Tropylium cation   Triphenyl
                     carbocation   methane

The triphenylmethyl cations are among the most stable carbocations known. They are intensely coloured and are readily formed when the corresponding triaryl carbinol is dissolved in strong acid.

$(C_6H_5)_3C - OH$ $\xrightleftharpoons{H_2SO_4}$ $(C_6H_5)_3C - \overset{\oplus}{O}H_2$ $\xrightleftharpoons{- H_2O}$ $(C_6H_5)_3\overset{\oplus}{C}$

Triphenylcarbinol   Triphenylmethyl cation
(colourless)   (orange yellow)

Several equivalent structures can be written for the tropylium cation so that only one-seventh of the positive charge is expected to be on each carbon. Since the cation also has six π electrons, it is anticipated to be unusually stable for a carbocation.

Tropylium ion
(hybrid structure)

The infrared and Raman spectra of tropylium bromide in hydrobromic acid solution have no common band. This shows that the cation exists in a highly symmetrical form in this solution.

$\overset{\oplus}{C_7H_7}$ + $\overset{\ominus}{OH}$ $\rightleftharpoons$ C_7H_7OH

The equilibrium constant for the above reaction is such that the cation is half converted to the carbinol at about pH 5.

□□□

14

Fluxional Molecules

In case of bicyclo[5,1,0]octadiene (also known as homotropilidene) (I), the initial and final structures (a and b) of a Cope rearrangement (3,3-sigmatropic rearrangement of 1,5-diene) may be identical. Molecules undergoing the reversible degenerate **Cope rearrangement** are referred to as *"fluxional molecules"*.

(a)	(I)	(b)
Bicyclo[5,1,0]octadiene		Homotropilidene

Homotropilidene may be prepared by cuprous chloride catalysed addition of diazomethane to cycloheptatriene.

Cycloheptatriene   Diazomethane   Homotropilidene

The NMR spectrum of homotropilidene at –50°C show signals of cyclopropyl, allylic and vinylic protons. As the temperature rises, the cyclopropyl signal is lost, the vinyl proton signal is much reduced and the allylic protons are merged into a broad signal. At 180°C, the spectrum is again resolved with the development of a four proton signal at τ 6.7 representing an average for protons between an allylic and a vinylic situation. This change is attributed to a very rapid rearrangement between (I), (a) and (b). So, homotropilidene is described as a fluxional molecule, since at relatively low temperature, its structure is in a state of flux between the equivalent structures.

© The Author(s) 2023
V. K. Ahluwalia and R. Aggarwal, *Alicyclic Chemistry*,
https://doi.org/10.1007/978-3-031-36068-8_14

There are many other derivatives of homotropilidene which exhibit fluxional property. Among them, the four 2,6-bridged compounds are semibullvalene (a), barbaralane (b), dihydrobullvalene (c) and bullvalene (d).

(a)   (b)   (c)   (d)

14.1   BULLVALENE

Bullvalene (II) is a fluxional molecule that can be converted into equivalent structures in about 10^6 ways, *e.g.*,

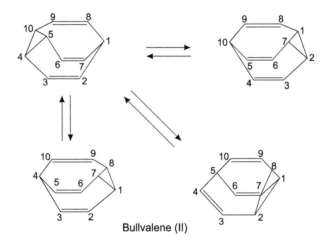

Bullvalene (II)

Rapid interconversion between these valence isomers gives a time-average structure in which all the protons become equivalent. The NMR spectrum at 100° shows a strong broad signal at τ 5.8, but at –25°C, vinylic (6H) at τ 4.3 and allylic (4H) protons at τ 7.9 can be recognised (Doering et al., 1963, 1967; Schrodel et al., 1965).

Bullvalene is obtained from cyclooctatetraene by thermal dimerisation followed by photolytic cleavage.

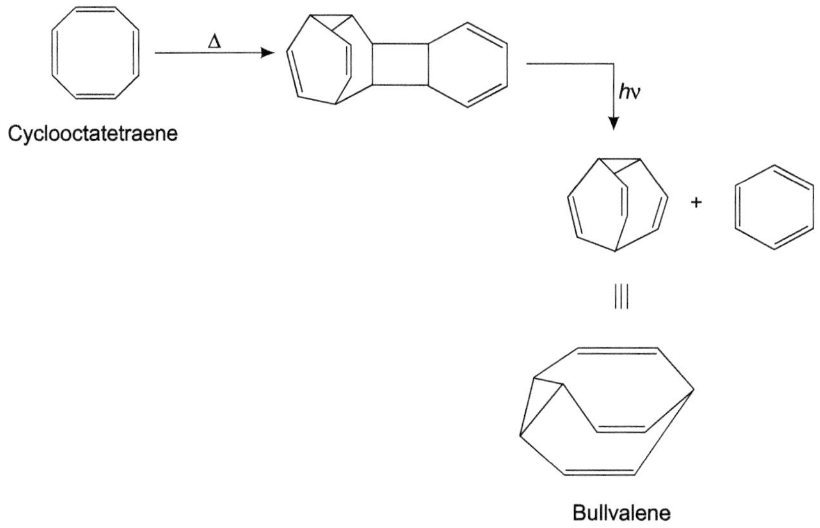
Δ
hν
Cyclooctatetraene
+
Bullvalene

❑❑❑

Catenanes, Rotaxanes and Knots

15.1 INTRODUCTION

Catenanes and rotaxanes differ from all other organic compounds synthesised in that the molecular subunits are linked mechanically without the aid of a chemical bond. In catenanes, the open chain compounds are made to cyclise. During the process, it may so happen that after a molecule has cyclised, another open chain compound passes through the cyclised ring and then cyclises to form a system of interlocked rings as shown below:

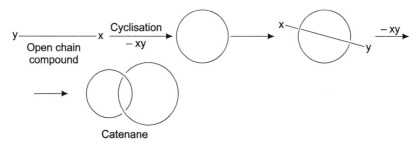

Pathway for the formation of interlocked rings (Catenanes)

However, the possibility of formation of interlocked rings is extremely low.

In rotaxanes, an open chain compound is made to cyclise in the presence of another open chain compound having bulky end groups. At some stage, the open chain compound may cyclise in such a way that the other open chain compound may pass through the closed ring. Such compounds are called rotaxanes (In Latin, *rota*: the wheel, axis: the axle) and have a ring like a wheel on an axle. In this case, the bulky end groups prevent the extrusion of the threaded chain from a macrocycle. The formation of a rotaxane is represented in the following diagram:

© The Author(s) 2023
V. K. Ahluwalia and R. Aggarwal, *Alicyclic Chemistry*,
https://doi.org/10.1007/978-3-031-36068-8_15

Pathway for the formation of a ring having a threaded chain (a rotaxane)

As in the case of catenanes, the possibility of formation of compounds like rotaxane is extremely low.

In knots, a single long chain may cyclise in such a way so as to form a compound with a knot like look. Its formation may be visualised as shown below. For the present, the discussion is centered only around catenanes and rotaxanes.

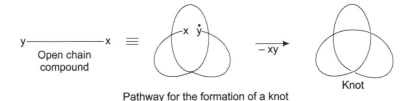

Pathway for the formation of a knot

15.2 NOMENCLATURE OF CATENANES AND ROTAXANES

A number of ways have been suggested for the nomenclature of catenanes.

(*i*) Frisch and Washerman (1961) used the designation 34,34-catenane for a compound consisting of two 34-membered macrocycles.

(*ii*) Subsequently, Tauber (1969) named a catenane (structure shown below) as [34,34]-catenane-1,18,1′,18′-tetraone -1,1′ dioxime; however, this nomenclature created confusion.

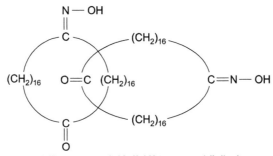

[34,34]-catenane-1,18,1′;18′-tetraone-1,l′ dioxime

(*iii*) Kohler and Dieterich suggested the word 'cum' (Latin - with) to be inserted between the individual components of a catenane. So, the above structure was named "cyclotetratriacontane-1,18-dione-1-oxime cum cyclotetratriacontane-1,18-dione-1-oxime".

A better system for the nomenclature was developed by Schill (1969). According to this system, the above structure was named catenane from cyclotetratriacontane-1,18-dione-1-oxime   and   cyclotetratriacontane-1,18-dione-1-oxime". Rotaxanes were also similarly named. The number of molecular subentities is given in brackets at the beginning of the name. Thus, the simplest catenane is [2]-catenane. The brackets are followed by the names of the molecular entities which are also enclosed in brackets. The rings of a catenane are numbered as in the case of normal or branched paraffins; the number of each is placed ahead of its name.

In catenanes and rotaxanes containing multiple windings, it is necessary to designate the winding number α. The following examples will clarify the above system of nomenclature:

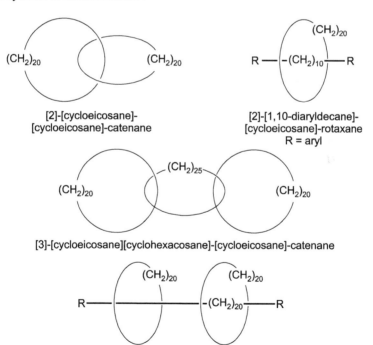

[2]-[cycloeicosane]-
[cycloeicosane]-catenane

[2]-[1,10-diaryldecane]-
[cycloeicosane]-rotaxane
R = aryl

[3]-[cycloeicosane][cyclohexacosane]-[cycloeicosane]-catenane

[3]-[1,20]-diaryl-eicosane]-[cycloeicosane]-(cycloeicosane)-rotaxane R = aryl

The present discussion is centered only around the catenanes and rotaxanes.

15.3 CATENANES

These are large ring compounds consisting of interlocked rings. Synthesis of catenanes by the use of acyloin condensation has been reported. In this procedure, long-chain dicarboxylic esters are converted to large ring compounds with high dilution technique; the method is the best for closing rings of ten numbers or more.

$$n = 10 - 20$$

Acyloin (yields 60 – 90% for 10 – 20 membered rings)

Acyloin condensation

A compound containing an interlocking ring (a catenane) can be obtained by acyloin condensation using a ester of 34-carbon dicarboxylic acid.

34–carbon acyloin   catenane (yield <1%)

Evidence for the formation of catenanes has been obtained by carrying out ring closure with the ester of 34-carbon dicarboxylic acid in the presence of deuterated 34-carbon cycloalkane. A product (catenane) was obtained with 68 carbons which contained one acyloin group and deuterium; its IR spectra showed v_{C-D} bands.

Ester of 34-carbon -dicarboxylic acid   Deuterated 34-carbon cycloalkane   Catenane (68-carbon compound)

The structures of deuterated catenane were confirmed by its oxidation to release the deuterated hydrocarbon.

The completely reduced catenane (a hydrocarbon) in which both C = O and CHOH groups are reduced to CH$_2$ is represented as follows:

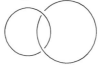

Representation of completely reduced catenane

The catenanes have great prospect as a polymer. However, the yield is very low. It is, in fact, a chance product depending upon the possibility of threading of diester molecule through the acyloin ring before it closed.

15.4 ROTAXANES

Various statistical methods are available for the synthesis of rotaxanes. Here, we are describing only one method in which a long chain(b) is bonded to bulky end groups (a) in the presence of macrocycle (c).

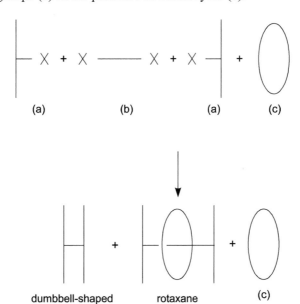

This scheme for the synthesis of rotaxane was investigated by Harison et al. They prepared the rotaxane by reaction of 2-hydroxycyclotria-contanone (1) with succinic anhydride which produced half ester (2). The sodium salt of the half ester was reacted with the chloromethylated copolymers from styene and divinylbenzene to form resin bond macrocycle (3). The resulting resin bonded macrocycle was reacted with

1,10-decanediol and triphenylmethyl chloride in the presence of pyridine, dimethylformamide and toluene. Hydrolysis with sodium bicarbonate in refluxing methanol then produced 6% of rotaxane (4).

The structure of rotaxane was confirmed by infrared spectroscopy and chemical degradation. Oxidation with silver oxide produced octacosane-1, 28-dicarboxylic acid as dimethyl ester with dumbbell-shaped molecule (5).

The cleavage of rotaxane (4) with BF_3-etherate in benzene gave decane-1, 10-diol, triphenylmethanol and acyloin (1).

Multiple Choice Questions

1. The cycloalkane, [structure: cyclohexane with CH₃, —CH₂CH₃, and Cl substituents] in named as

    (a) 1-chloro-3-ethyl-4-methylcyclohexane

    (b) 4-chloro-2-ethyl-1-methylcyclohexane

    (c) 1-ethyl-2-methyl-3-chlorocyclohexane

    (d) 1-methyl-2-ethyl-4-chlorocyclohexane

2. The bicyclic compound [structure] is named as

    (a) Bicyclo [3.1.0]hexane   (b) Bicyclo [1.3.0] hexane

    (c) Bicyclo [0.1.3] hexane   (d) Any of the above name

3. The reaction of 1,3-dibropropane with zinc gives cyclopropane. This is an example as:

    (a) Wurtz reaction   (b) In termolecular wurtz reaction

    (c) Intramolecular wurtz reaction (d) None of the above.

4. Which of the following starttment is/are correct about cycloalkanes

    (a) In general the bps and m.p's of cycloalkanes are higher than those of alanes

    (b) Cycloalkenes unlike alkanes are soluble in water

    (c) Cycloalkanes are transparent to uv light above 220 mm

    (d) At low temperature cyclohexane exhibts two signals.

5. The heat of combustion per CH_2 group in cyclopropane (I) cyclobutane (II), cyclopentane (III) and cyclohexane (IV) is of the order.

    (a) I > II > III > IV   (b) IV > III > II > I

    (c) II > III > IV > I   (d) I > III > II > IV

6. Which at the following has least ring strain.

    (a) Cyclobutane   (b) Cyclopentane

    (c) Cyclohexane   (d) Cyclooctane

© The Author(s) 2023
V. K. Ahluwalia and R. Aggarwal, *Alicyclic Chemistry*,
https://doi.org/10.1007/978-3-031-36068-8

7. The product obtained in the following reaction is

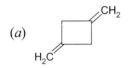

(c) A mixture of (a) and (b)   (d) There is no reaction

8. The major product obtained in the following reaction is

$$CH_2 = C = CH_2 \xrightarrow{400°C} ?$$

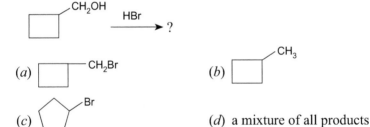

(c) A mixture of (a) and (b)

(d) The allene molecule gives CO_2 and H_2O

9. The product obtained in the following reaction is

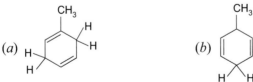

(c) (d) a mixture of all products

10. The reaction of toluine with Na in liquid NH_3 using alcohol as solvent gives.

(a)   (b)

(c) a mixture of (a) and (b)   (d) There is no reaction

11. Thermal dimensation of butadiene the major product as

(a) 4-Vingylcyclohexene   (b) 1, 2-divinylcyclobutane

(c) a mixture of (a) and (b)   (d) There is no reaction

12. At room temperature, in case of *tert.* butylcyclohexane the major form is in which *tert*-buty l group is presentin

(*a*) axial position   (*b*) equatorial position

(*c*) both forms are present in 1 : 1 ratio

13. Which of the following statements are correct incase of disubstituted cyclohexanes

(*a*) *trans* diequatorial conformation of 1,2-dimethyl cyclohexane is more stable than the *trans*-diaxial cyclohexane

(*b*) In case of 1,3-dimethyl cyclohexane, the *cis* form with diequatorial substituent is more stable than the *trans* form

(*c*) In case of 1,4-dimethylcyclohexane, the *trans* form with diequaterial substitution is more stable than the *cis* form.

(*d*) all are correct.

14. In case of cholistrol

(*a*) Rings A and B are *cis* fused

(*b*) Rings B and C are *trans* fused

(*c*) Rings C and D are *trans* fused

(*d*) all are correct

15. Which of the following cycloalkenes exist only in the *cis* form

(*a*) Cyclopropene   (*b*) Cyclobutene

(*c*) Cyclopentene   (*d*) All the above cycloalkenes

16. The alkene $\begin{array}{c}H_3C\\H\end{array}C=C\begin{array}{c}CH_3\\C-CH_3\\\parallel\\O\end{array}$ can be epoxidised by

(*a*) CH_3COOOH (*b*) $C_6H_5\text{-}COOOH$

(*c*) alkaline H_2O_2 (*d*) all of the avove

17. The acid catalysed ring opening of isobutylene oxide with methanol gives

(*a*) $\begin{array}{c}H_3C\\H_3C\end{array}C\begin{array}{c}OCH_3\\-CH_2\\\mid\\OH\end{array}$ (*b*) $H_3C-\underset{\underset{OH}{\mid}}{\overset{\overset{CH_3}{\mid}}{C}}-CH_2OCH_3$

(*c*) a mixture of (*a*) and (*b*)   (*d*) there is no reaction

18. The following reaction gives

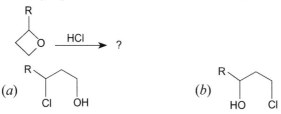

(a) R—C̤—C̤—CH₃ with O—O—CH₃, H on top, R, CH₃ below

(b) R—C̤—C̤—H with O—C—CH₃, CH₃ on top, R', CH₃ below

(c) a mixture of (a) and (b)   (d) there is no reaction

19. The major product obtained in the following reaction is

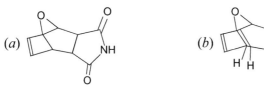

(a)   (b)

(c) a 1 : 1 mixture of (a) and (b)   (d) there is no reaction

20. The reactivity of furan (I), pyrole (II) and thiophlene (III) is of the order

(a) I > II > III   (b) III > II > I

(c) II > I > III   (d) III > I > II

21. Furan can be converted into THF by

(a) H₂/Pt/HOAc   (b) Na – C₂H₅OH

(c) Zn/HOAc   (d) Rane Ni

22. Diels-Alder reaction of furan with maleimide gives

(a)   (b)

(c) Mixture of (a) and (b)   (d) There is no reaction

23. The reacion of pyrrole with chloroform – NaOH gives

(a)   (b)

(c) a mixture at (a) and (b)   (d) There is no reaction

24. Thiophen can be nitrated give 2-nitrothiophen by using
    (*a*) Conc. HNO_3 (*b*) dilute HNO_3
    (*c*) Conc. HNO_3/conc. H_2SO_4 (*d*) Conc. $HNO_3 + Ac_2O$ at 10°C

25. Indole can be converted into 2,3-dihydroindole by reduction using
    (*a*) Raney Ni, C_2H_5OH, 100°C, 85 atm
    (*b*) H_2/Ni
    (*c*) Li-NH_3, NaOH
    (*d*) all the above reducing agents

26. The basicity of pyridine (I), pyrrole (II) and aliphalic amins (III) is it the order
    (*a*) I > II > III   (*b*) III > II > I
    (*c*) III > I > II   (*d*) I > III > II

27. Arrange piperidine (I), pyridine (II) and pyrrole (III) in order of their basic stringth
    (*a*) II > I > III   (*b*) I > II > III
    (*c*) III > I > II   (*d*) I > III > II

28. Quinoline can be reduced to give 5,6,7,8-tetrahydro quinoline using
    (*a*) Sn/HCl   (*b*) H_2/PtO_2/CF_3COOH
    (*c*) Ni/N_2/210°/70 atms   (*d*) H_2/Ni/210°

29. Isoquinaline can be converted into phthalimide by
    (*a*) Oxidation with alkaline $KMnO_4$
    (*b*) Oxidation with neutral $KMnO_4$
    (*c*) Treatment with CH_3CO_3H
    (*d*) Treatment with cone HNO_3

30. Cyclobutadiene is
    (*a*) aromatic   (*b*) antiaromatic
    (*c*) non aromatic

31. Which of the following is aromatic
    (*a*) cyclopentane   (*b*) cyclopentadienyl cation
    (*c*) cycloheptatriene   (*d*) cycloheptatrienyl cation

ANWERS

1.(b)	2.(a)	3.(c)	4.(a)(c)(d)	5.(a)	6.(c)	7.(a)	8.(b)
9.(c)	10.(a)	11.(a)	12.(b)	13.(d)	14.(d)	15.(d)	16.(c)
17.(a)	18.(c)	19.(b)	20.(a)	21.(d)	22.(a)	23.(c)	24.(d)
25.(a)	26.(c)	27.(b)	28.(b)	29.(b)	30.(b)	31.(b)(d)	

Fill in the Blanks

1. The molecular formula of cycloalkanes is ....

2. The cycloalkanes are called ....

3. The IUPAC name of ⬡—$CH_2CH_2CH_2CH_3$ is ....

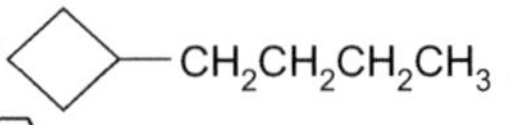

4. The compound ⬡CH_2 is named as ....

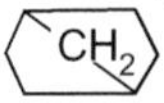

5. The compound ⬡—CH_3 is named as ....
   H_3C

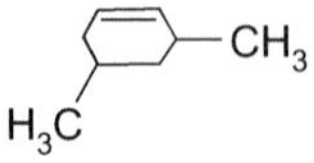

6. The reaction of 1, 6-dibromohexane with zinc gives.

7. The reaction of ethyl acetoacetate with 1,4-dibromebutane in presence of sodium ethoxdie gives ....

8. Intra molecular claisen condensation of ethyl adipate in presence of NaOEt to give 2-carbethoxycyclpentanone is known as ....

9. Of the various cycloalkanes, ... has the lowest heat of combustion per CH_2 group.

10. The reaction of 2-chlorocyclobutane with NaOMe gives cyclopropane carboxylic and this is an example of ....

11. The product in the following reaction is................... .
    CH_3 CH_3
    $\;\;\;\;$C—CH_2Cl $\xrightarrow{\bar{P}h\overset{+}{N}a}$?
    CH_3

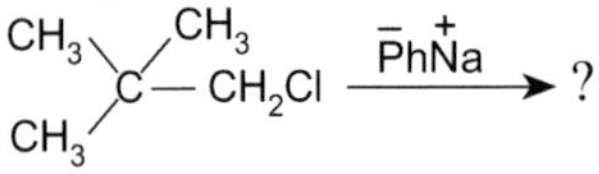

12. The product obtained in the following reaction is
    Me $\;\;\;\;$ CH_2Cl
    $\;\;\;\;$C=C $\xrightarrow{Buli}$?
    Me $\;\;\;\;$ Me

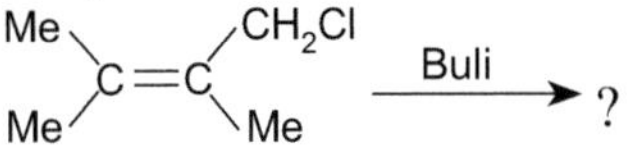

13. The reaction of ketene with didzomethane gives ....

14. The reaction of 2-bromocyclohexanone with sodium methoxide gives ....

15. Out of the two conformations of cyclohexane (chair or boat), the most stable conformation is ....

16. Cyctooctene can exist in .... form.

17. The reaction of $\underset{H}{\overset{Ph}{\diagdown}}$C=C$\underset{Ph}{\overset{H}{\diagup}}$ with oxone and acetone gives ....

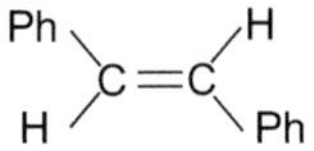

18. The following convessies

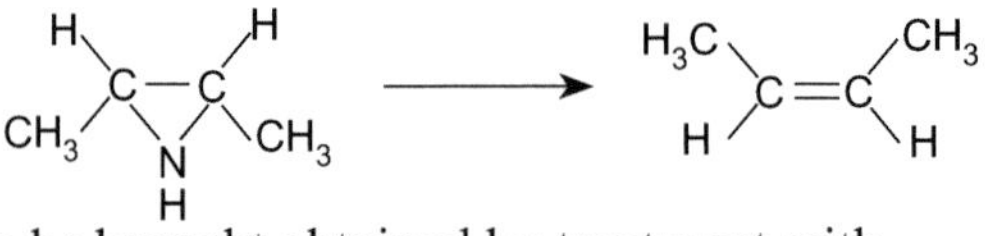

    $\underset{CH_3}{\overset{H}{\diagdown}}$C—C$\underset{CH_3}{\overset{H}{\diagup}}$ over N—H $\longrightarrow$ $\underset{H}{\overset{H_3C}{\diagdown}}$C=C$\underset{H}{\overset{CH_3}{\diagup}}$

    can be brought obtained by treatment with ....

19. The reaction of 2-phenyl oxetane with HCl gives ....

20. The following reaction gives .... $\xrightarrow{\text{HNO}_3}$?

21. Pyrrole on treatment with ethyl didzoacetate at 100° in presence of Cu gives ....

22. The reaction of pyrrole with chlorobenzene in presence of KNH_2 gives ....

23. Dichlorocarbene on reaction with 3-methyl indole gives ....

24. Quriolive can be converted into pyrdine – 2, 3 dicarboxaldehyle by ......

25. Azulene on heating (> 350°C) gives ....

26. The following structure in designated as ....

27. The reaction of benzine with diazomethane under photochemical conditions gives ....

ANSWERS

1. C_nH_{2n}

2. Carbocyclic compounds

3. 1-Cyclobutylpentane

4. Bicyclo [2.2.1]heptane

5. 3,5-Dimethylcyclohexene

6. Cyclohexane

7. Cyclohexylmethyl ketone

8. Dieckmann condensation

9. Cyclohexane

10. Favorskii rearrangement

11.

12.

13. Cyclobutanone

14. Methyl cyclopentane carboxylate

15. *cis* conformation

16. both *cis* and *trans*.

17.

18. NOCl

19.

20.

21.

22. 2-Phenylpyrrole

23. 3-Chloro-2-methyl quirioline

24. Oxidation with O_3 and treatment with Me_2S.

25. Naphthalene

26. Tricycto [3, 3.1.1$^{3.7}$] decane

27. Tropilidene

Fill in the Blanks

1. Give IUPAC names to the following:

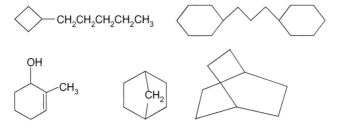

2. Give an example of intromolecular wurtz reaction.

3. Starting with diethyl molanate how will you synthesies cyclopropane, cyclobutane and cyclopentane.

4. Write a note an Blanc's rule.

5. Give an example of intramolecular claisen condensation.

6. Write notes on

   (*a*) Thorpe Zieglers method.   (*b*) Acyloin condinsation.

7. Which signal you will observe in the NMR spectra of cyclohexane at ordinary tempeations and at very low temperature.

8. Write a note an Baeyers strain theory.

9. Discuss the relative stabilities of cycloalkanes.

10. Explain Favorskii rearrangement giving one example.

11. How is diazomethane obtained.

12. Write a note on Simmon's reaction and Wagner-Meerein rearrangement.

13. Discuss the conformations of cyclobutane and cyclopentane.

14. Discuss the conformations 1,2-,1,3- and 1,4- disubstituted cyclohexanes. In each case, which confermation in more stable.

15. How will you establish that cyclooctatetraene exists as bicyclo [4.2.0]-2,4,7 - octatriene.

16. Cyclopropene, cyclobutene and cyclopentene exists only in the *cis*-form, comment ?

17. Discuss the importane of heterocyclic compounds.

18. How is dimethyl dioxirane obtained. Give a use of this compound.

19. Starting with isobutyline oxide how with you obtain

$$H_3C \underset{H_3C}{\overset{OCH_3}{\diagup}} C - CH_2 \quad \text{and} \quad H_3C - \overset{CH_3}{\underset{OH}{\overset{|}{C}}} - CH_2OCH_3$$
$$\underset{OH}{|}$$

20. Write notes on:

    (a) Palerno – Buchi reaction

    (b) Pal - Knorr synthesis

    (c) Fiest - Benary synthesis

    (d) Henrzch Pyrrole synthesis

    (e) Madelung indole synthesis

21. Discuss the relative reactively of furam, pyrrole and thiophen.

22. How will you demonstrate that Furam undergoes electrophilic substitution at position?

23. Give a synthesis of 2-bromofuram.

24. Why nucleophilic substitutions are not normally encontered in pyrrole.

25. Explain what is cine substitution.

26. How in indole synthesised.

27. How in indole converted into indigo.

28. Nitration of pyridins gives 3-nitro and 2-nitro pyridine. How is 4-nitro pyrdine obtained.

29. How is 3-bromopyridine converted into 3-aminopyridine.

30. Discuss the mechanism of skraup synthesis for making qurnaline.

31. How is the numbering of isoquinoline done.

32. Explain the terms Aromatic, antraromatic and non-aromatic.

33. How would the relative position of keto group and double bond be established in civetone?

34. Give one synthesis of muscone.

35. Azulenes have intense blue-violet colour. Discuss.

36. Give the names of the following fused ring systems.

37. Discuss the mechanism of conversion of dimer of cyclopentadiene into adamantane.

38. What are fluxional molecules? Discuss giving example of bullvalene.

39. Discuss the aromatic behaviour of tropolone.

Index

© The Author(s) 2023

V. K. Ahluwalia and R. Aggarwal, *Alicyclic Chemistry*,

https://doi.org/10.1007/978-3-031-36068-8

❑❑❑